Soumaya Haouel

Essential oil: Extraction processes & Biological activities

Soumaya Haouel

Essential oil: Extraction processes & Biological activities

ScienciaScripts

Imprint

Any brand names and product names mentioned in this book are subject to trademark, brand or patent protection and are trademarks or registered trademarks of their respective holders. The use of brand names, product names, common names, trade names, product descriptions etc. even without a particular marking in this work is in no way to be construed to mean that such names may be regarded as unrestricted in respect of trademark and brand protection legislation and could thus be used by anyone.

Cover image: www.ingimage.com

This book is a translation from the original published under ISBN 978-613-8-46821-9.

Publisher:
Sciencia Scripts
is a trademark of
Dodo Books Indian Ocean Ltd., member of the OmniScriptum S.R.L Publishing group
str. A.Russo 15, of. 61, Chisinau-2068, Republic of Moldova Europe
Printed at: see last page
ISBN: 978-620-4-08725-2

Table of contents:

Effect of extraction processes on the yield, chemical composition , insecticidal activity and antioxidant power of *Eucalyptus salubris* F. Muell essential oils.

Authors:

Soumaya Haouel Hamdi

Marwa Ben Jemia

Jouda Mediouni Ben Jemâa

Summary

The objective of this work is to study the impact of different extraction processes on the yield, chemical composition, antioxidant activity and insecticidal power of *Eucalyptus salubris* essential oils against *Sitophilus oryzae* adults.

Three extraction processes of essential oils were tested: a conventional or traditional method the simple hydrodistillation (HS) and two new or modern methods the extraction assisted by ultrasound (EAU) and the extraction assisted by microwave (EAM). In addition, the collections were made in two arboerta of the region of Gabes El- hamma and Zrig. The chemical composition of these oils was determined by chromatographic analyses. These analyses revealed that the oils of the arboretum of El-hamma are rich in hydrocarbon monoterpenes (49.2-69.9%) while those of the arboretum of Zrig are rich in oxygenated monoterpenes (27.7-39.5%) mainly 1,8-cineole, in addition to the presence of two other major compounds which are: a-aminoanthraquinone (19.7%- 28.9%) and cycloisolongifolene (27.8-35.2%) In addition, the results showed the presence of both quantitative and qualitative differences between the extraction processes.

The insecticidal power of these essential oils was evaluated by fumigation tests against adults of *S. oryzae*. The results showed that the fumigant power depends on the region of harvest, the extraction process, the concentration and the exposure time. The conventional method (HS) gave the best performance for both arboreta in terms of mortality and lethal concentrations. For modern methods, ultrasound assisted extraction (UEE) was better for essential oils extracted from *E. salubris* leaves of El-hamma (LC50: 34.43gl/l air) than those of Zrig (LC50: 137.75gl/l air). In contrast, microwave-assisted extraction (MAE) was better for essential oils from Zrig (LC50: 117.09^l/l air) than that from El-hamma (LC50: 46.737^l/l air).

The results of the antioxidant activity showed that the *E. salubris* oils tested have a significant antioxidant activity that varies with the region of collection as well as the extraction processes. The highest inhibitions were obtained for Zrig oils with the following ranking: HS (PI: 69.25%) > EAM (PI: 65.32%)> WATER (PI: 58.27%). In contrast, for the essential oils of El-hamma, the modern processes EAU and EAM had the highest inhibition percentages (PI: 59.79% and PI: 56.58% respectively) compared to the conventional HS method (PI: 42.28%).

Keywords: Essential oils; *Eucalyptus salubris*; simple hydrodistillation; ultrasound assisted extraction; microwave assisted extraction; *Sitophilus oryzae*.

Abstract

The purpose of this work is to study the impact of different extraction processes on the yield, chemical composition, antioxidant potential and insecticidal activity of *Eucalyptus salubris* essential oils against adults of *Sitophilus oryzae*.

Three extraction process of essential oils have been tested: a conventional or traditional method Simple Hydrodistillation (HS) and two new or modern methods Ultrasonic Assisted Extraction (EAU) and Microwave Assisted Extraction (EAM). In addition, plant material collections were carried out in two arboerta of the region of Gabès El-hamma and Zrig.

The chemical composition of these oils was determined by chromatographic analyzes. These analyzes revealed that the oils from El-hamma arboretum were rich in hydrocarbon monoterpenes whereas those of the Zrig arboretum were rich in oxygenated monoterpenes, mainly 1,8-cineole, in addition to their high levels of a-aminoanthraquinone and cycloisolongifolene. Moreover, the results showed the presence of quantitative rather than qualitative differences between the extraction processes.

The insecticidal potential of these oils was assessed by fumigation on *S. oryzae* adults. The results showed that fumigant toxicity depends on the harvest area, extraction process, concentration and exposure duration. The conventional method (HS) achieved the best performances for both arboreta in terms of insect mortality and lethal concentrations. For modern methods, ultrasonic assisted extraction (UAE) was better for the oils harvested from El-hamma (LC50: 34,43 pl/l air) than those from Zrig (LC50: 137,75 pl/l air). In contrast, Microwave Aided Extraction (EAM) was better for Zrig's harvested oils (LC50: 117,09 pl/l air) than for El-hamma (LC50: 46,737 pl/l air).

The results of the antioxidant activity showed that *E. Salubris* tested oils have an important antioxidant activity which varies with the harvest area as well as the extraction processes. The most important inhibitions were obtained for Zrig oils with the following classification order: HS (69.25%)> EAM (65.32%)> EAU (58.27%). In contrast, for oils of El-hamma, modern processes UAE and EAM accomplished the highest inhibition percentages (PI: 59.79% and PI: 56.58% respectively) compared to the conventional method HS (PI: 42.28%).

Keywords : Essential oils, *Eucalyptus salubris*, Hydrodistillation simple, Ultrasonic Assisted Extraction, Microwave Assisted Extraction, *Sitophilus oryzae*.

الملخص

تأثير أساليب الاستخراج على المردود والتركيبة الكيميائية، والفعالية الحشرية والقوة المضادة للأكسدة للزيوت الروحية من‏*EucalyptusSalubris*

يهدف هذا العمل إلى دراسة تأثير طرق الاستخراج على المردود والتركيبة الكيميائية، والنشاط المضاد للأكسدة والنشاط الحشري من الزيوت الروحية من‏*Eucalyptus Salubris*‏ضد الطور البالغ لحشرة سوسة الأرز ‏*Sytophilus oryzae*.

تم اختبار ثلاث طرق لاستخراج الزيوت الروحية: طريقة تقليدية التقطير المائي العادي (HS)‏طريقتين جديدتين او حديثتين و هما الاستخراج بمساعدة‏الموجات فوق الصوتية ‏(EAU)‏والاستخراج بمساعدة الميكروويف (EAM).

بالإضافة إلى ذلك، تم جمع العينات النباتية من مشجرين بمنطقة قابس الحامة وزريق. تم تحديد التركيبة الكيميائية لهذه الزيوت عن طريق التحليل الكروماتوغرافي. وكشفت هذه التحاليل أن الزيوت من مشجر الحامة غنية‏بمونوتربان الهيدروكربونية الا ان زيوت مشجر زريق غنية بمونوتربانالأوكسيجين وخاصة 8،1 سينيول، بالإضافة إلى محتويات عالية من α-aminoanthraquinone و‏cycloisolongifolène. إضافة إلى ذلك، أظهرت النتائج وجود فروق كمية وغير نوعية بين طرق الاستخراج.

تم تقييم الفعالية الحشرية لهذه الزيوت عن طريق اختبارات التبخير ضد حشرات الطور البالغ لـ ‏*S. oryzae*. وأظهرت النتائج أن الفعالية الحشرية تختلف حسب منطقة الجني، وطريقة الاستخراج، والتركيز ومدة التعرض. أعطت الطريقة التقليدية ‏(HS)‏أفضل أداء لكل من المشجرين من حيث الوفيات و قيمة التركيزات القاتلة. بالنسبة للطرق الحديثة، كانت طريقة الاستخراج بمساعدة الموجات فوق الصوتية‏(EAU)‏أفضل للزيوت مشجر الحامة ‏(CL_{50} :34,43 ميكرو لتر/لتر هواء) من زيوت مشجر الزريق‏CL_{50} :137,75 ميكرو لتر/لتر هواء)، على خلاف ذلك، كانت طريقة الاستخراج بمساعدة الموجات الدقيقة ‏(EAM)‏ أفضل لزيوت مشجر زريق‏CL_{50}117.09 (ميكرو لتر/لتر هواء)‏مقارنة بمشجر الحامة ‏(CL_{50} :46,737‏ميكرو لتر/لتر هواء).

بينت نتائج النشاط المضاد للأكسدة ان زيوت ‏*E. Salubris*‏ لها نشاط هام مضاد للأكسدة ويختلف حسب منطقة الجني وطريقة الاستخراج. وقد تم الحصول على أهم نشاط لزيوت‏مشجر زريق‏مع التصنيف التالي‏‏HS (PI:"69.25%)> : ‏EAM(PI:65.32%)%. و خلاف ذلك، بالنسبة لمشجر الحامة، كانت طرق الاستخراج الحديثة لها أكبر نسبة نشاط (59.79% و56.58% على التوالي) مقارنة مع الطريقة التقليدية (42.28%).

كلمات مفاتيح: زيوت روحية، ‏*Eucalyptus Salubris*‏ استخراج بمساعدة‏الموجات فوق الصوتية، التقطير المائي العادي، ‏استخراج بمساعدة الميكروويف، الصوتية،‏*Sitophilus oryzae*.

PCA: Principal Component Analysis.

AFNOR: Association française de normalisation.

LC50: median lethal concentration that kills 50% of the population.

LC95: median lethal concentration that kills 95% of the population.

GPC: Gas chromatography.

***E. salubris** :Eucalyptus salubris.*

EAM : microwave assisted extraction.

WATER: extraction assisted by ultrasound.

GCMS: Gas chromatography coupled with mass spectrometry.

EO: essential oil.

HS: simple hydrodistillation.

MC: adjusted mortality.

PI: percentage of inhibition.

S. oryzae: *Sitophilus oryzae.*

iil/l air: Micro litre per litre of air.

General introduction

Cereals are the main sources of human and animal nutrition in the world. Wheat ranks first in world production and second, after rice, as a source of food for human populations (Bajji, 1999). By virtue of the size of the areas occupied and its role in the country's food security, the cereal sector remains one of the most important components of agricultural production in Tunisia (Cheikh, 2004). The area sown annually is on average around 1.45 million hectares (Bachta, 2011). During their conservation, cereal grains are progressively damaged quantitatively and qualitatively, the losses are estimated at 100 million tons of which 13 million are caused by insects. The post-harvest losses of cereals in developed countries are around 3% and in Africa they reach 30% (Silvy, 1992).

The pests of stored products form a group in which Coleoptera and Lepidoptera predominate (Lepigre, 1951). Among the most damaging beetles, the rice weevil *Sitophilus oryzae* (Linnaeus, 1763) (*Curculionidae*), is universally recognized as one of the most devastating pests of stored grain, not only because of its own consumption, but also because it opens the door to a whole range of detritus feeders, the most common of which is *Tribilium castaneum* (Herbst, 1797), which completes the damage (Markham, 1994).

The responses to combat insects, the main pests of stocks, have been mainly chemical. However, given the nuisances associated with the use of pesticides, selection of resistant strains, environmental pollution, intoxication, the search for alternatives is necessary (Momar et *al.*, 2010). Thus, substances of plant origin, in particular essential oils have received particular interest given their interesting insecticidal potential, their antioxidant potential, their biodegradability and the broad spectrum of target insects as well as the various modes of action with which they are equipped.

It is in this perspective that the present work was carried out to evaluate the insecticidal potential by fumigation of essential oils of *Eucalyptus salubris* (F. Muell, 1876) collected from arboreta of El-hamma and Zrig and extracted by the conventional HS (simple hydrodistillation) and modern EAU (ultrasound assisted extraction) and EAM (microwave assisted extraction) processes against adults of *Sitophilus oryzae*.

The approach adopted for this work involved the study of the following aspects:

- Rearing of *S. oryzae* under controlled laboratory conditions,

Extraction of *E. salubris* oils by different processes,

Determination of the yield and chemical composition of essential oils,

Evaluation of the insecticidal activities of essential oils by fumigation against adults of *S. oryzae*,

Evaluation of the antioxidant activity of these oils obtained by the different processes.

This study is organized in three parts:

The first part contains bibliographical data on the plant and the insect pest studied,

The second part describes the different methods and techniques used,

The third part contains the results and discussions.

Part I: Bibliographic synthesis

1. The *Myrtaceae* family :

The *Myrtaceae* family is a family of dicotyledonous plants that includes about 3800 species divided into 133 genera (Oldrich, 2005).they are trees and shrubs, often producers of aromatic oils of temperate, subtropical to tropical areas, growing mainly in Australia, tropical America, Mediterranean region, sub-Saharan Africa, Madagascar, tropical and temperate Asia, and the Pacific islands. In this family, we can mention the genera: *Eucalyptus, Psidium of* which the guava tree is a part, *Myrtus of* which the myrtle tree is a part (shrub of the Mediterranean scrub), *Eugenia of* which the clove tree (*Eugenia cariophyllata*) which gives the clove (Flemming, 1997).

2. Eucalyptus

2.1. Origin

The genus *Eucalyptus* (family *Myrtaceae*), is one of the most important and most planted genera in the world, and is represented by about 700 species. In 1957, Tunisia proceeded to the introduction of 117 species of *Eucalyptus* which were planted in more than 30 arboretums, distributed throughout the country. It is mainly used for the production of wood mines and in the fight against soil erosion (Khouja et *al.*, 2001; Ben Marzoug et *al.*, 2010).

2.2. Classification

Kingdom: Plant

Phylum: Spermaphytes

Subphylum : Angiosperms

Class: Dicotyledons

Subclass: Dialypetals

Family: Myrtaceae

Genus: Eucalyptus

2.3. General characteristics of *Eucalyptus*

The *Eucalyptus is* a large exotic tree. It can reach gigantic proportions, up to 150 metres in height in its country of origin. In the Mediterranean basin, it reaches 30 meters (Metro, 1970; Ben Hassine, 2008).

Most *Eucalyptus trees* are evergreen but some tropical species lose their leaves at the end of the dry season. Like other members of the *Myrtaceae* family, *Eucalyptus* leaves are covered with oil glands. Abundant oil production is an important characteristic of this genus (Bakkali el *al.*, 2008). *Eucalyptus trees* require acidic and nutrient rich soil. They are very sensitive to cold (Ben Hassine, 2008).

2.4. Description of the species *Eucalyptus salubris* (F. Muell, 1876)

E. salubris is known as the typical gimlet gum. This species is widely distributed in inland areas of south-western Australia. It is a medium-sized tree reaching 20-25 m in height with a short fat and spreading crown. The bark is smooth and copper-coloured on a spiral fluted trunk 0.75 m in diameter. The young leaves are alternate, petiolate and narrowly lanceolate, while the older leaves are alternate, narrowly lanceolate and short petiolate (Jacobs, 1982). *E. salubris* is distinguished by an abundant flowering spread over more than 6 months (Khouja et *al.*, 2006). This species is suitable for soil conservation, windbreaks and ornamental plantations in the dry zone. The tree is small in size for fuelwood production (Jacobs, 1982) (Figure 1).

Figure 1 :Eucalyptus salubris (anonymous3)

3. Essential oils

3.1. Definition

Essential oils are volatile, fragrant plant extracts, also known as liquid aromatic organic substances, found naturally in various parts of trees, plants and spices, and are volatile and sensitive to the effect of heat (Evans, 1998).

According to the [4th] edition of the French pharmacopoeia (2000), essential oils are products of generally rather complex composition, containing volatile principles. The AFNOR standard NF T 75-006 (2000) defines essential oils as "products obtained either from natural plant materials by steam distillation, or by mechanical processes from the epicarp of citrus fruits, or by dry distillation. The essential oil thus obtained is separated from the aqueous phase by physical processes'.

According to the 7th edition of the European Pharmacopoeia, essential oils are defined as, "An odoriferous product, usually of a complex composition, obtained from a plant raw material, either by steam driving, dry distillation or by a suitable mechanical method without heating. An essential oil is usually separated from the aqueous phase by a physical process that will not lead to significant changes in its chemical composition" (Asbahani et *al.*, 2015).

Essential oils can be extracted by various methods. Due to their hydrophobic nature and their density often lower than that of water, essential oils are generally lipophilic, soluble in organic solvents and immiscible with water. They can be separated from the aqueous phase by decantation (mechanical separation). However, the extraction yields vary according to the species and organs. They remain very low (about 1%), which makes them very rare and valuable substances. Among plant species, only 10% contain essential oils and are called aromatic plants (Svoboda and Greenaway, 2003).

3.2. Location in the plant

Essential oils exist practically only in aromatic and medicinal plants. The most known genera producing essential oils are distributed in a limited number of families: *Myrtaceae, Rutaceae, Laminaceae, Asteraceae, Apiaceae, Cupressaceae, Poaceae, Zingiberaceae, Piperaceae* (Belyagoubi, 2005).

Essential oils can be stored in all vegetative organs: leaves, flowers, bark, wood, roots, rhizomes, fruits and grains. Synthesis and accumulation are generally associated with the presence of specialized histological structures (Belyagoubi, 2005). The organs of synthesis of essential oils are often located near the surface of the plant. They are essential oil cells in the *Lauraceae* or *Zingiberaceae*, secretory hairs in the Lamiaceae, secretory pockets in the *Myrtaceae* or *Rutaceae*, and secretory ducts in the *Apiaceae* or *Asteraceae* (Bruneton, 1993).

3.3. The different methods of extraction of essential oils

Extraction seeks to isolate a complex plant mixture, constituents or compounds with different chemical properties (Peyron, 1992). There are several techniques for extracting essential oils. These techniques do not present the same degree of effectiveness with regard to, on the one hand, very volatile and not very polar molecules, and on the other hand share of low volatile and highly polar molecules (Asbahani et *al.*, 2015). According to Asbahani etal. (2015), essential oils are obtained from plant material by several extraction methods.

3.3.1. Traditional or conventional extraction methods

3.3.1.1. Hydrodistillation

The principle corresponds to a heterogeneous distillation. The process consists of immersing the raw plant material in a still filled with water placed on a hot plate. This mixture is brought to a boil, generally at atmospheric pressure. The heat allows the odorant molecules contained in the plant cells to be broken down and released. It is then cooled in a cooler and condensed in an essencier or Florentine vase. Once condensed, water and aromatic molecules, due to their differences in density, separate into an aqueous phase and an organic phase (the essential oil). In the laboratory, the system equipped with a cohobe which is generally used for the extraction of essential oils in accordance with the European pharmacopoeia is the Clevenger. This method has the advantage of being simple and efficient (Henry et *al.*, 2014) (Figure 2).

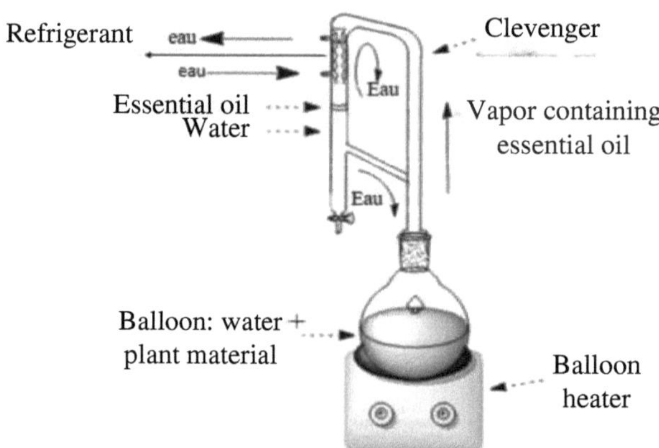

Figure 2: Operating principle of the Hydrodistillation apparatus (Mahtab et al.,2017).

3.3.1.2. Steam training

Steam distillation is one of the essential methods for obtaining essential oils. Unlike hydrodistillation, this technique does not bring water and the plant material to be treated into direct contact. The steam supplied by a boiler passes through the plant material located above a grid. As the steam passes through the material, the cells burst and release the essential oil, which is vaporized under the water.

the action of heat to form a "water + essential oil" mixture. The mixture is then conveyed to the condenser and essencer before being separated into an aqueous phase and an organic phase: the essential oil (Gawde et *al.*, 2014) (Figure 3).

Figure 3: Steam extraction (own photo)

3.3.2. Modern methods or green extractions

Modern extraction methods are also called green extractions. These techniques are beneficial to the environment and aim to "design chemical products and processes to reduce or eliminate the use and synthesis of hazardous substances" (Gérin, 2002). This concept of green extraction was made popular in the scientific community by the American chemists Paul Anastas and John Warner, thanks to their publication in 1998 of the book "Green Chemistry: Theory and Application" in which they present the twelve principles of green chemistry and advocate the reduction, recycling or elimination of hazardous substances that are harmful to humans and the environment (Anastas, 1998)

3.3.2.1 . Microwave assisted extraction

In this process, the plant material is heated by microwave in a closed chamber in which the pressure is sequentially reduced. The volatile compounds are carried away by the steam. They are then recovered using conventional condensation, cooling and decantation processes (El haib, 2011) (Figure 4).

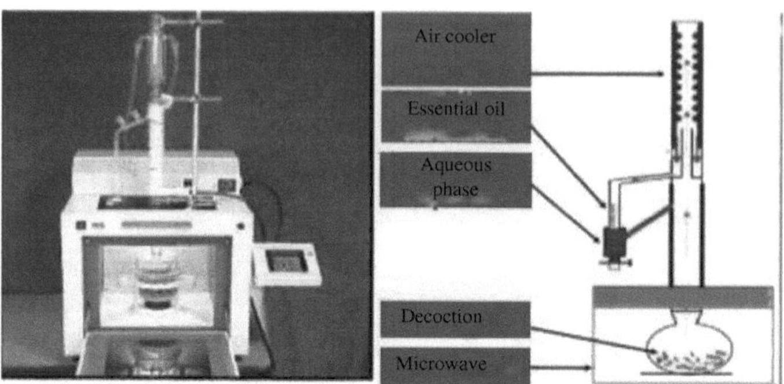

Figure 4: Setup of a microwave-assisted extraction (El haib, 2011).

The main advantage of this technique is that it considerably reduces distillation time (extraction time divided by 5 to 10 times) and energy (lower temperature).

Microwaves or microwave frequencies are electromagnetic waves with wavelengths intermediate between infrared and broadcasting waves between 300 GHz and 300 MHz. However, only a few frequencies are used for industrial, scientific and medical (ISM) purposes between 0.915 and 2.45 GHz. At the latter frequency, matter reacts with the electromagnetic field mainly *via* the phenomenon of dipole rotation and induced polarization (Leonelli et *al.*, 2013).

3.3.1.3. . Ultrasonic or sonication assisted extraction

This technique represents an adaptation of hydrodistillation or organic solvent extraction. Indeed, the raw material is immersed in water or in the solvent, and at the same time it is subjected to the action of ultrasounds. The technique can be used for the extraction of EO, but it has been developed mainly for the extraction of certain molecules of therapeutic interest (De castro and Capote, 2007) (Figure 5).

Ultrasound refers to mechanical and elastic sound waves with frequencies higher than those detectable by the human ear, i.e. around 18 kHz. Sound waves generate mechanical vibrations in a solid, liquid or gas. Unlike electromagnetic waves, sound waves can propagate through a material and involve cycles of expansion and compression as they propagate through the medium. Expansion can create bubbles that form, expand and collapse in a liquid. Near a solid surface, cavity collapse is asymmetric and produces a high velocity liquid jet (Benamor, 2008).

Figure 5: Ultrasonic assisted extraction process in the laboratory (Boutamani, 2013).

3.3.2.3 Supercritical fluid extraction

The supercritical fluid extraction technique has attracted the interest of many researchers and has been successfully used in environmental, pharmaceutical, food and industrial analysis. All substances on earth have three states: solid, liquid and gas, while the supercritical state is distinct and can only be reached if the substance is subjected to a temperature and pressure above its critical point.

Carbon dioxide (CO_2) is considered the ideal solvent for supercritical extraction, its critical temperature is 31°C which is close to room temperature, while its critical pressure is low in the order of 74 bar. The disadvantage of CO_2 is that it has a low polarity which makes it ideal for lipids, fats and non-polar substances, but inappropriate for most pharmaceutical samples (Azmir et *al.*, 2013).

3.3.2.4 Superheated water extraction

In trying to find alternative methods to replace organic solvents in the extraction of natural products from plants, a technique using superheated water has been used for some decades. The objective was to reduce pollution in the workplace and the environment and to avoid the undesirable residues of organic solvents that are often present in food products and perfumes (Clifford, 2002).

Superheated or subcritical water is liquid water above its boiling temperature (100°C) under pressure. Under these conditions, liquid water is less polar than at room temperature and has an increasing ability to dissolve organic compounds, thus giving it an extended character of polar organic solvents. The term superheated water is attributed to liquid water under pressure at a temperature between 100°C and 374°C which is its critical temperature. For the extraction of natural products, the temperature cannot be too high because of their thermo-sensitivity, so the range between 100 and 200°C is the most used (Clifford, 2002).

Compounds with high solubility in superheated water are generally polarizable, such as aromatic compounds, essential oils or those with a polar character (Clifford, 2002). The advantages of this technique include low cost, non-toxic, higher yield, good quality clean products, environmental benefit, and energy savings (Clifford, 2002).

3.4. Chemical composition

Chemically, essential oils are highly volatile molecules with an extremely complex structure, synthesized from methyl-2-buta-1,3-diene (isoprene) units (Hernandez Ochoa, 2005). They can contain several hundred different chemical molecules. The most frequently encountered are alcohols (phenols and sesquiterpenols), ketones, terpene aldehydes, esters, ethers, terpenes and oxides (Anonymous, 2008).

The chemical composition varies according to environmental conditions, genotype, geographical origin, harvesting period, drying, drying location, temperature and duration of drying, presence of parasites, viruses and weeds (Mohammedi, 2006). The most important changes occur during hydrodistillation under the influence of operating conditions, especially the environment (pH, temperature) and the duration of extraction (Hernandez Ochoa., 2005). Other factors such as treatments undergone before or during hydrodistillation (grinding, dilaceration, chemical degradation, pressure, agitation) contribute to the variation of the composition of the essential oil (Hernandez Ochoa, 2005). This author reported that variations in the chemical composition of essential oils from the same plant can be attributed to hybridization, genetic polymorphism or mutations. Other factors can influence the chemical composition of essential oils, such as the organ used and the vegetative cycle (Hernandez Ochoa, 2005).

The main groups of chemical constituents of essential oils are the terpene group and the group of aromatic compounds derived from phenylpropane. Among the chemical compounds frequently found are the fatty acid degradation compounds and the terpene degradation compounds. All these terpene degradation compounds are formed during photosynthesis (Garneau, 2001).

Eucalyptus oil is a complex mixture of varieties of monoterpenes and sesquiterpenes and aromatic phenols, oxides, ethers, alcohols, esters, aldehydes and ketones, the exact composition and proportion of which vary according to the species(Brooker and kleinig, 2006).

3.5. Biological activities of essential oils

3.5.2. Antioxidant activity

Oxidation remains a major problem that affects the quality of food, leading not only to a decrease in the nutritional value of the food, but also to effects that are known to be harmful to the consumer and that may be associated with cancer risks in humans (Bandoniene et *al.*, 2000).

Oxidizing substances can have different chemical structures such as proteins, unsaturated fatty acids, cholesterol, phospholipids and mainly lipids which, because of their establishment, are very sensitive to oxidation (El Kalamouni, 2010). In addition, most foods contain traces of metals resulting from packaging or storage or even from the food composition (Labuza, 1971).

It is impossible to control virtually all the factors that cause oxidation, so the presence of antioxidants in the diet has become essential for food quality and safety. Indeed, the antioxidant is a substance that, in small quantities, responsible for slowing the oxidation of easily oxidizable compounds such as fatty acids (Frankel et *al.*, 2000).

The many natural properties of essential oils make them both nutraceutical ingredients and very promising preservatives for the food industry. Each essential oil has a specific activity that varies according to environmental conditions. However, the use of essential oils is a relevant choice when faced with a specific contamination risk or the need to reduce or replace chemical or synthetic preservatives. Also, their use in very small quantities is possible because of their great effectiveness, unlike some additives such as salts or spices. In addition, the addition of essential oils to a food could give it a nutraceutical value (Caillet and Lacroix, 2007).

3.5.3. Antifungal and antimicrobial activity

Eucalyptus essential oils and their main constituents possess toxicity against a wide range of microbes, including pathogenic bacteria and fungi. Previous work has shown that essential oils reduce mycelial growth (Fiori et *al.*, 2000), inhibit spore production and germination (Fiori et *al.*, 2000).

3.5.4. Herbicidal activity

The essential oil extracted from *Eucalyptus* species shows phytotoxicity against weeds and has great potential for weed management (Batish et *al.*, 2007).

3.5.5. Acaricidal activity

Essential oils and their bioactive compounds can be effectively used to dispel ticks and mites (Saad et *al.*, 2006).

3.5.6. Insecticidal activity

Eucalyptus essential oil can act directly as a natural insect repellent to protect against mosquitoes and other arthropod pests or to combat herbivores (Daizy et *al.*, 2008). The essential oils of plants belonging to the genera *Eucalyptus*, have demonstrated their insecticidal effectiveness (Tapondjou et *al.*, 2002).

The fumigation toxicity of *Eucalyptus* essential oils has been studied on various species of beetles and demonstrated that the toxic effects depend on the insect species and the time of exposure to the essential oil. In this context, Mediouni-Ben Jemâa et *al.* (2012) reported the toxicity of *Eucalyptus* essential oil on date insect pest.

The pesticidal activity of *Eucalyptus* essential oils is due to its compounds such as 1,8 cineole, citronellal, citronellol, citronellyl acetate, p-cymene, eucamalol, limonene, linalool, alpha-pinene, gamma terpinene, alpha terpineol, alloocimene and aromadendrene. (Batish et *al.*, 2008).

The various compounds in Eucalyptus essential oils are synergistic in providing overall pesticidal activity (Cimanga et *al.*, 2002). Among the various compounds in *Eucalyptus* oil, 1,8-cineole or Eucalyptol is the most important. In fact, this compound is characteristic of the genus *Eucalyptus* and is largely responsible for its various pesticidal properties (Duke, 2004).

3.5.6.1 Rice weevil: *Sitophilus oryzae* (L.)

The rice weevil, *Sitophilus oryzae* is an important pest of rice, wheat, oats, rye and maize in the world (Sabbour, 2012). In Tunisia, Jerraya (2003) reported that it is a very damaging insect attacking several host commodities. It is a small (2 to 4 mm), elongated, flat-backed insect, easily recognizable because its head is extended by a "rostrum" perfectly visible to the naked eye. Its elytra are decorated with four reddish spots).

3.5.6.2 Systematic position

The rice weevil or *S.oryzae has* the following systematic position (Lepesme, 1944):

Branching :	Arthropods
Class:	Insects
Order :	Beetles
Family:	Curculionidae
Genre:	*Sitophilus*
Species:	*Sitophilus oryzae* (Linnee, 1763)

3.5.5.3 Origin and geographical distribution

According to Codon & William (1991), *S. oryzae* is a species found mainly in tropical and subtropical areas, although the country of origin of this species is the Indian region. At present, this insect is cosmopolitan and is distributed throughout the world because of international trade. In the tropics, *S. oryzae is* often confused with a closely related species, *Sitophilus zeamais* (Motschulsky,

1855), which is larger and more particularly depreciates maize (Codon & William, 1991).

3.5.5.4. Life cycle

The development of *S. oryzae* is holometabolous type i.e. from larva to nymph and then adult (the larva is very different from the adult). The larva is apodal (legless), worm-like (worm-like in appearance) and the pupa is immobile (Aragon, 2014). The development of the insect takes place inside the grain. With her rostrum, the female makes a hole in a grain, lays an egg and then fills the hole with mucilage that will harden in contact with air. As soon as it appears, the larva digs a gallery through the grain which it will enlarge as it grows. It will then transform into a nymph in the cavity it has prepared and, after a final moult, will become an adult which will then emerge from the grain to reproduce (jerraya, 2003). During this totally hidden development, the interior of the grain will have been consumed entirely.

An emptied grain with a hole in it containing the excrement of the larval development will then remain (Figure 6). Under optimal conditions (25±5°C temperature and 70% relative humidity), the development cycle lasts about one month and the adult can live up to 5 months (Benazzeddine, 2010).

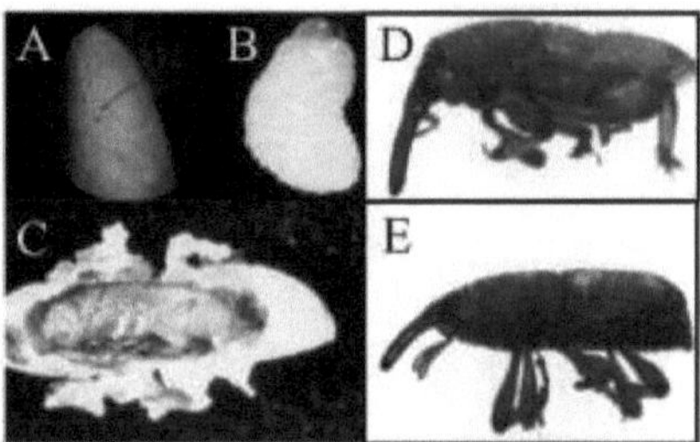

Figure 6: The different developmental stages of Sitophilus oryzae

(A: egg; B: larva; C: nymph; D and E: adult) (A: Benazzeddine., 2010; B: anonymous1 C : anonymous2; D and E: personal photograph)

3.5.5.5. Damage

The rice weevil feeds and multiplies at the expense of many cereals: rye, oats, barley, wheat, rice, maize ... (Cruz and Troude, 1988) and can even attack chickpea and cowpea (Egbon and Ayertey, 2013). The damage of *S. oryzaesis* mainly caused by the larvae which can consume half to one third of the endosperm of a wheat grain (Balachawsky, 1963). Rice weevil is one of the primary pests for stored cereals on which it causes decrease in seed weight, deterioration in quality by promoting fungal development (Kranz et *al.*, 1977); thus facilitating the development of secondary pests (De-Groot, 2004) (Figure 7).

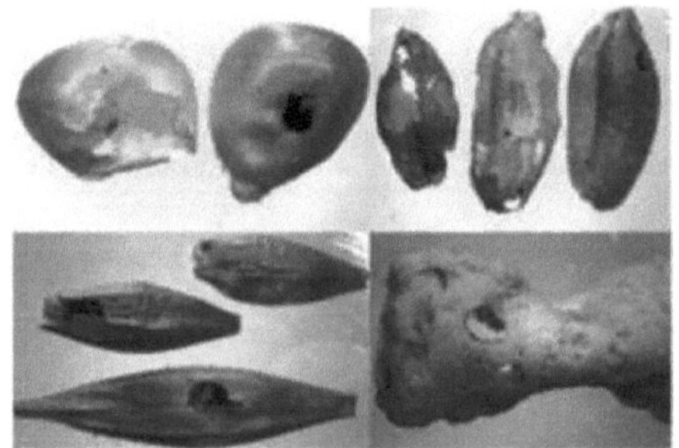

Figure 7: Damage of Sitophilus oryzae (own picture)

1.5.5.6. Control methods for the rice weevil

The protection of stored foodstuffs from attack by this pest involves a range of different methods and techniques.

1.5.5.6.1. Chemical control

Chemical control is based on the use of synthetic organic compounds. Synthetic insecticides are classified into organic insecticides (organochlorines, organophosphates, carbamates) which represent the majority of insecticides currently used, and inorganic insecticides which are generally based on arsenic or lead and which have a severe toxicological impact, hence the ban on their use in many countries (Regnault-Roger and Philogene, 2005). Two groups of products can be used: contact insecticides and fumigants:

- Contact insecticides:

Contact treatment consists in covering the grains as well as the storage places with the insecticide which acts on the pest. Insecticide abuse and misuse is often very common in storage systems and represents a major problem (Benayad, 2008).

- Fumigants:

Fumigation has been used as a method of pest control for over 100 years (Taylor, 1994). Fumigation of stored commodities involves treating the products with a toxic gas. This gas, called fumigant, kills insects if it is kept in contact with the grain for a sufficiently long period of time at a certain concentration. The main advantage of fumigation is the ability of the gas to penetrate the grain and thus destroy the eggs, larvae and pupae that develop there (Jerraya, 2003).

1.5.5.6.2. Physical Wrestling

- Cold control: This involves lowering the temperature in storage areas to 10°C in order to slow down the development of the insect (Armitage, 1987).
- Heat control: The technique consists in treating the grains with temperatures higher than 60°C. The rise in temperature ensures total mortality of the insect without altering the quality of the product (Troude et *al.*, 1988).
- Mechanical control: Adult weevils are removed by shaking and sieving. These techniques are not effective against the hidden forms (eggs and larvae) (Hebert et *al.*, 1972).

1.5.5.6.3. Biological control

Biological control of *S. oryzae relies on the* use of its natural enemies whether they are predators or parasitoids to control this damaging pest and keep it below the economic damage threshold (Tietze and Modi, 2000). The most common are Hymenoptera of the family Pteromalidae (Benazzeddine, 2010).

On the other hand, in recent years, the use of plant extracts and mainly essential oils has been booming due to the multitude of pesticidal properties they have. In this context, essential oils extracted from various plant species have shown in addition to their therapeutic and antimicrobial properties, very interesting insecticidal potential against several pests of stored products (Bakkali et *al.* , 2008).

Part II: Materials and Methods

1. Plant material

1.1. Sampling

The sampling was carried out in the South of the country, in the region of Gabes in arboreta of El-hamma and Zrig belonging to the National Institute of Research in Rural Engineering, Water and Forests (INRGREF). The collection of *Eucalyptussalubris* leaves was carried out in May 2015.

Table 1: Characteristics of the two collection Arboreta

The arboretum	El-hamma	Zrig
Creation	April 1960 from species of Australian origin	
Governorate	Gabes	
Geographical coordinates	Latitude 33 deg 54' 32.05"N, Longitude 9 deg 39' 47.26"E	Latitude33deg43 58.08"N, Longitude 10 deg 09' 33.80"E
Altitude (m)	42	45
The bioclimatic floor	Upper Saharan	Lower arid

Table 2: Soil characteristics in the two Arboreta (INGREF)

Arboretum	pH	EC (mS/cm)	CaCO3 (%)	Gypsum (%)	Texture		
					Sand (%)	Clay (%)	Silt (%)
El-hamma	7,9	9,7	5,5	1,3	1,8	5,9	1,8
Zrig	7,9	0,9	7,23	0,25	89,8	6,9	1,0

EC: Conductivity; CaCO3: active limestone

1.2. Drying of foliage

The harvested leaves are dried in the shade, protected from light, for 7 days at an ambient temperature of about 25°C. The dried plant material is then cut into small pieces of 4 to 5 cm.

2. Extraction of essential oils

The extractions were carried out at the Laboratory of Applied Fluid Mechanics, Process Engineering and Environment at the National Engineering School of Sfax (ENIS).

The extraction of essential oils is carried out by three methods: a conventional method by Simple Hydrodistillation (HS) and two modern methods: Extraction assisted by ultrasound (EAU) and Extraction without solvent assisted by microwaves (EAM).

2.1. Extraction by Simple Hydrodistillation :

The extraction of essential oils is carried out using a Clevenger-type hydro-distiller (Figure 8). The latter consists of a chamber where the plant material is emerged in boiling water. The water vapor formed is rich in essential oils, it passes through the coil that surmounts the chamber. The water is then condensed as soon as it reaches the refrigerator. Based on the low density of essential oils compared to water, two phases are obtained by decantation:

An organic product that contains only essential oils,

The other is aqueous, which contains some aromatic compounds, generally of higher density than water.

Hydrodistillation takes 4 hours. The essential oils are recovered and stored at a temperature of -4°C protected from light until the moment of use.

Figure 8: Simple Hydrodistillation Process (Own shot)

2.2. Ultrasonic assisted extraction :

This process is similar to Hydrodistillation except that the mixture is heated with ultrasonic waves (Figure 10). In this study, 80g of *Eucalyptus* leaves were immersed in distilled water in a 1 litre capacity chamber in a Clevenger system and placed in an ultrasonic bath (InterSonic, UB-405 model, internal dimensions: 500x285x255 mm).

Figure 9: Ultrasonic bath (Own shot)

2.3. Microwave assisted extraction

Microwave-assisted hydrodistillation is similar to conventional hydrodistillation except that the mixture is heated with microwaves. The experimental protocol is as follows: 80g of *Eucalyptus* leaves are immersed in water in a reactor placed in the microwave oven chamber (Figure 11). The refrigeration system and the part intended for the recovery of the essences are located outside the oven.

The heating releases the essential oil which is then carried away by the steam produced from the water in the plant material. A cooling system outside the microwave oven allows the condensation of the distillate, composed of water and essential oil.

Figure 10: Microwave extractor (Own shot)

3. The parameters studied :

3.1. The yields of essential oils

The yield of essential oil is the ratio of the weight of the extracted oil (in g) to the weight of the sample used. The yield expressed as a percentage (%) is calculated by the formula of Mohammedi (2006):

With : R : oil yield in %.

       PB : Weight of the essential oil in grams

       PA: weight of plant material in grams

3.2. Chromatographic analysis of essential oils:

The chromatographic analyses of the essential oils of the *E. salubris* species were carried out at the Laboratory of Biological Engineering of the National Institute of Applied Sciences and Technology (INSAT).

The essential oil was analysed using a combination of gas chromatography-mass spectrometry (GC-MS) methods by means of the Agilent- technologies 6890 N GC apparatus (Figure 12). This system is equipped with a flame ionization detector, split mode injector and HP-5MS capillary column (30 m*0.25 mm, 0.25 pm film thickness; Agilent-technologies, Little Falls, CA, USA). The injector and detector temperatures were set at 220°C and 290°C, respectively. The column temperature was programmed from 80°C to 220°C at a rate of 4°C/min, with lower and upper temperatures being held for 3 and 10 minutes, respectively. The flow rate of the carrier gas (Helium) was 1.0ml/minute. A 1pl sample was injected in three replicates, using split mode. All quantifications were performed using a built-in data processing program provided by the chromatograph manufacturer. Composition was reported as a relative percentage of the total peak area. The identification of essential oil constituents was based on a comparison of their n-alkane retention times with published data and spectra of authentic compounds. The compounds were defined and authenticated using their mass spectra against the Wiley library version 7.0.

Figure 11: GC-MS device (INSAT) (Own picture)

3.3. Antioxidant activity of essential oils

The antioxidant activity of *E. salubris* essential oils was carried out at the Laboratory of Plant Ecology at the Faculty of Sciences of Tunis (FST).

The DPPH (2,2-diphenyl-1-picrylhydrazyl) radical presents, at room temperature, an intense violet coloration which disappears on contact with a bioactive molecule capable of saturating its electronic

$$\text{DPPH}^{\bullet} + \text{AH} \rightarrow \text{DPPH} - \text{H} + \text{A}^{\bullet}$$

layer according to the following reaction

(violet)   (Colorless)

The detection of the decrease in the intensity of this coloration is done by spectrophotometry. The reduction of DPPH highlights the antiradical power of the tested plant extract.

A 250 |il test portion of the essential oil is placed in the presence of 1 ml of an ethanolic solution of DPPH. The mixture is allowed to stand for 30 minutes in the dark for incubation, after which the absorbance is measured at 517 nm using a spectrophotometer against a control (no extract). The results are expressed as a percentage of inhibition calculated following the decrease in the intensity of the coloration of the mixture according to the following equation.

Anemone: Absorbance of the control.

Aextract: Absorbance of the extract.

3.4. Insecticidal activity of essential oils

3.4.1. Breeding of *Sitophilus oryzae*

The rice weevil *S. oryzae* used in the bioassays is reared at the Laboratoire de Biotechnologie Appliquée à l'Agriculture (Section Entomologie) of the Institut National de la Recherche Agronomique de Tunisie (INRAT).

Rearing was done in plastic boxes containing either rice or maize. The rearing was maintained at a temperature of 25±1°C and a relative humidity of 45%, with a photoperiod of 12 hours light/12 hours dark (Figure 12).

Figure 12: Rearing of Sitophilus oryzae (own photo)

3.4.2. Toxicity by fumigation :

The fumigant activity of the essential oils was tested on 7-10 day old adults of *S. oryzae*. The insects were exposed to four increasing doses of oils: 1, 2, 3 and 6pl. The oils were applied to 2 cm diameter filter paper discs (Whatman paper). These doses are converted into concentrations calculated in relation to the volume of air in the spittoon and expressed in microlitres per litre of air (pl/l air). Thus, the doses tested correspond to the following concentrations: 16.67; 33.33; 50 and 100 pl/l air. Adults were placed in 60 ml spittoons with 10 adults (males and females) in the presence of 20 g of corn. The spittoons were then sealed. For each concentration, the test was repeated 3 times. Untreated individuals serving as controls were maintained under the same conditions. Mortality was assessed every 6 hours until 100% mortality was obtained. An individual is considered dead when it shows no sign of life (completely immobile) (Figure 13)

Figure 13: Fumigation device (Own shot)

The evaluation of the efficacy of essential oils against adults of *S. oryzae* is calculated using Abbot's formula (Abbot, 1925) expressed by the following formula:

Where MC: Adjusted mortality (%)

C: Control mortality (%)

T: Mortality in treated patients (%)

3.4.3. Lethal Concentrations

The effectiveness of the essential oils is determined by the respective values of the median lethal concentrations (LC50 and LC95). The LC50 lethal concentrations correspond to the concentrations of oils that induce 50% mortality in the treated insects, and the LC95 lethal concentrations correspond to the concentrations of oils that induce 95% mortality in the treated insects. To determine them, 10 individuals are placed in spittoons and exposed to the different oil concentrations already used for the fumigation tests. Mortality was assessed after 24 hours of exposure. LC50 values are calculated using the probit analysis described by Finney (1971).

3.4.4. Statistical analysis

The analysis of the results was done using SPSS version 20 software. The collected data were subjected to statistical analysis of variance by testing the different variables. In the presence of significant effect, analysis of variance (one factor ANOVA) was followed by ranking of means using Duncan's test. The collected data of essential oil yields were statistically investigated for correlation between yield and soil conditions. In order to identify the major compounds of the oils obtained from the leaves of *Eucalyptus salubris* species, the chemical composition and extraction processes were subjected to a principal component analysis (PCA). To assess the significance of the variability of the activity of the essential oils, the average mortality of *S. oryzae* adults as well as the percentages of the major compounds of the oils studied were subjected to analysis of variance and hierarchical clustering (HC). The data collected were statistically correlated with *S. oryzae* adult mortality, harvesting regions, extraction methods, concentration and exposure time.

Part III: Results and Discussion

1. Yield and chemical composition of essential oils

1.1. Yield of essential oils

The results of determining the yields of essential oils using three extraction methods from two harvesting regions are shown in Figure 14.

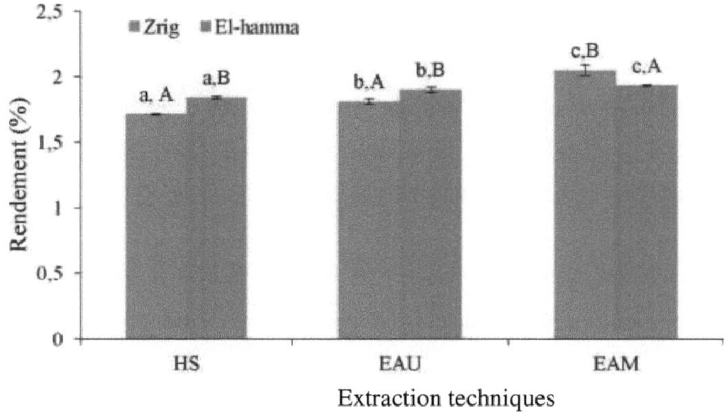

Figure 14: Essential oil yield of E. salubris according to the region of harvest and the three extraction methods

(HS: Simple Hydrodistillation; UAE: Ultrasound Assisted Extraction; MAE: Microwave Assisted Extraction). (For each region, comparisons were made between extraction methods (lower case letter) (Duncan test). For each extraction method, comparisons were made between harvesting regions (upper case letters) (Student's test)].

The analysis of these results shows that the essential oil yield varies according to the region of harvest and the extraction method. The analysis of variance of essential oil yields shows a highly significant effect of harvesting region (F=10.52; p=0.007) and extraction method (F=149.76; p<0.001). For the two regions of El- hamma and Zrig, the highest yield was obtained for essential oils extracted by modern methods with 1.81% and 2.05% for EAU and EAM respectively for Zrig region, and 1.9 and 1.93% for EAU and EAM respectively for El-hamma region. Thus, according to our results, the ranking order of the extraction processes is as follows: microwave assisted hydrodistillation > ultra sound assisted hydrodistillation> simple hydrodistillation (EAM>EAU>HS). This result can be explained by the explosion produced at the level of the cell wall as a result of the sudden increase in the 35

The yields of essential oils from El-hamma are the highest for simple hydrodistillation and ultrasonic-assisted extraction (EAM). On the other hand, for simple hydrodistillation and ultrasonic assisted extraction, the yields of essential oils from El-hamma are the highest. While, for microwave-assisted extraction, the Zrig region shows a higher yield (Figure 14).

Considering the results obtained, it can be concluded that regardless of the region and the extraction methods, the yields of essential oils present considerably important values. Indeed, Ben Marzoug et *al.* (2010) reported that the yields of essential oils of *E. salubris* vary from 1.51 to 4.8%. The results of Haouel et *al.* (2017) showed that the yield of *E. salubris* essential oils varies significantly depending on the extraction methods. Indeed, Simple hydrodistillation showed that the lowest yield is 1.63% followed by ultrasound assisted extraction 1.69%, while the yield obtained by microwave assisted extraction gave the highest yield of about 1.75%. This variation in yield can be attributed to edaphic characteristics, ecological and climatic conditions (Ben Marzoug et *al.*, 2010). In this context, Batish et *al.* (2008) reported that the yield of essential oil depends on the species, season, location, climate, soil type, age of leaves, process used for drying the plant material and method of oil extraction.

Table 3 illustrates the correlation matrix between essential oil yield and soil characteristics. The observation of this table shows that the correlation coefficients taken 2 to 2 at the threshold of 0.01% are very strong. Indeed, the extraction methods by simple hydrodistillation and the extraction assisted by ultrasound are characterized by a very highly significant negative correlation with altitude, active limestone content (CaCO3), sand, clay and silt. Similarly, microwave assisted extraction is characterized by highly significant negative correlation with conductivity (EC) and gypsum contents. While a highly significant positive correlation was observed between the simple hydrodistillation and ultrasonic assisted extraction methods and conductivity (EC) and gypsum contents.

Table 3: Correlations between essential oil yield and soil characteristics in the two Arboretums

	Altitude	CE	CaCo3	Gypsum	Sand	Clay	Silt
HS	-0,99**	0,99**	-0,99**	0,99**	-0,99**	-0,99**	-0,99**
WATER	-0,94**	0,94**	-0,94**	0,94**	-0,94**	-0,94**	-0,94**
EAM	0,92*	-0,92*	0,92*	-0,92*	0,92*	0,91*	0,92*

The correlation is significant at the 0.05 level (two-tailed).
The correlation is significant at the 0.01 level (two-tailed).

2. Chemical composition of essential oils

2.1. Variation in the chemical composition of essential oils according to the region of harvest and the three extraction methods

The chromatographic analyses related to the determination of the chemical composition of essential oils of *E. salubris* leaves according to the two harvesting regions (El-Hamma and Zrig) and according to the extraction methods are recorded in Table 4 where the compounds are classified according to their retention time (RT). The identified compounds were grouped into chemical classes (hydrocarbon monoterpenes, oxygenated monoterpenes, oxygenated sesquiterpenes, ketones and others) and the total percentage of identified compounds is displayed at the end of each oil column (% Total).

The results showed that the chemical composition of the essential oils varied according to the harvesting region and the extraction method. Chromatographic analyses (GC and GC/MS) identified 33 compounds representing 90.81 to 100% of the total oil with variable contents depending on the harvesting region (Table 4).

The essential oils of *Eucalyptus salubris* of El-Hamma obtained by different extraction processes (HS, EAU, EAM) are characterized by the highest percentages of monoterpenes. These proportions vary according to the method of extraction it is 50.97% for simple hydrodistillation; 56.95% for microwave assisted extraction and 63.13% for ultrasound assisted extraction. The main compounds identified are hydrocarbon monoterpenes (49.24% - 62.29%) and are represented mainly by sabinene and P-cymene. The hydrocarbon sesquiterpenes (8.53-8.85%) occupy the third position and are represented mainly by aromadendrene. The ketones (2.85-3.3%) occupy the [4th] position and are represented only by cryptone. Oxygenated monoterpenes (0.84-1.73%) are mainly represented by phellandral. Oxygenated sesquiterpenes are absent except for the microwave assisted extraction where there is an appearance of dehydro-cis-a-copaene-8-ol with a percentage of 2.08%.Other compounds appeared with very important percentages (24.83-29.26%) consisting mainly of cycloisolongifolene.

Concerning the essential oils of *Eucalyptussalubris* de Zrig obtained by the different extraction processes (HS, EAU, EAM), they are characterized by high contents of a-aminoanthraquinone (19,71-28,93%) and cycloisolongifolene (27,87-35,27%). Oxygenated monoterpenes (27.78-39.59%) occupy the second position and are represented mainly by 1,8-Cineole. The hydrocarbon monoterpenes occupy the third place with contents varying from 4.45% (EAM) to 8.63% (HS) (Table 4).

Table 4: Chemical compositions of Eucalyptus salubris essential oils according to the region of harvest and the three extraction methods.

	TR		El-hamma Arboretum			Zrig Arboretum		
			HS	WA	EAM	HS	WAT	EAM
		Hydrocarbon monoterpenes	**49,24**	**62,29**	**55,99**	**8,63**	**5,61**	**4,45**
1	5,38	a-Thujene	1,6	1,6	1,43	0	0	0
2	5,56	a-Pinene	4,08	5,1	4,78	2,95	3,72	2,87
3	6,92	P-Myrcene	0,75	0,94	0,89	0	0	0
4	7,34	a-Phellandrene	0,43	0,54	0,49	0	0	0
5	7,95	P-Cymene	13,05	14,95	13,62	0	0	0
6	8,11	Sabinene	21,15	31,35	27,48	0	0	0
8	13,34	2-Carene	0,34	0	0	5,01	0	0
9	10,99	8-Terpinene	4,77	6,34	5,24	0,67	0	1,58
10	23,01	Cadinene	1,55	1,47	2,06	0	0	0
11	13,32	a-Terpinolene	0	0	0	0	1,89	0
12	11,57	3-carene	1,52	0	0	0	0	0
		Oxygenated monoterpenes	**1,73**	**0,84**	**0,96**	**39,59**	**31,25**	**27,78**
13	16,02	Phellandral	0,74	0,84	0,96	0	0	0
14	17,33	o-Cymene-5-ol	0,99	0	0	0	0	0
15	8,16	1,8-Cineole	0	0	0	30,94	28,99	25,69
16	10,81	Fenchylalcohol	0	0	0	1,04	0	0
17	12,51	Borneo	0	0	0	0	2,26	2,09
18	11,65	a-Phellandreneepoxid	0	0	0	5,35	0	0
19	12,28	Pinocarvone	0	0	0	2,26	0	0
		Hydrocarbon sesquiterpenes	**8,53**	**8,73**	**8,85**	**3,67**	**0**	**0**
20	25,23	a-Gurjunene	0,71	2,18	0	0,78	0	0
21	25,73	Aromadendrene	4,2	4,65	7,56	2,89	0	0
22	26,00	8-Gurjunenne	1,85	0	0	0	0	0
23	26,27	P-Guaiéne	0,65	0	0,78	0	0	0
24	26,82	P-Guajen	0,42	0	0,51	0	0	0
25	27,296	Calamene	0	1,9	0	0	0	0
26	27,789	Valencene	0,7	0	0	0	0	0
		Oxygenated sesquiterpenes	**0**	**0**	**2,08**	**0**	**0**	**0**
27	27,297	Dehydro-cis-a-copaene-8-ol	0	0	2,08	0	0	0
		Ketone	**3,24**	**3,3**	**2,85**	**0**	**4,83**	**3,57**
28	13,083	Cryptone	3,24	3,3	2,85	0	0	0
29	11,601	p-Mentha- 1,3,8-triene	0	0	0	0	4,83	3,57
		Other	**28,07**	**24,83**	**29,26**	**47,58**	**58,31**	**64,2**
30	14.89	Benzaldehyde	0,74	0	0,63	0	0	0
31	26.52	Methanonaphthalene	0,69	0	0	0	0	0
3231	.96	a-Aminoanthraquinone	5,38	2,95	3,69	19,71	27,46	28,93
3325	.61	Cycloisolongifolene	21,26	21,88	24,94	27,87	30,85	35,27

Total (%)	90,81	99,99	99,99	99,47	100	100

According to Garneau (2001), the chemical composition of plant EO varies according to the species, geographic and climatic factors, and annual or seasonal conditions. It could also vary from one individual to another. Indeed, Pavela (2009) reported that the chemotype of the plant, the cultivation practices and the method of extraction also intervene in the variation of the chemical composition of EO. Previous work conducted by Mansouri et *al.* (2011) indicated that the difference in the chemical composition of EOs observed between different species of aromatic plants could be explained by an adaptation to biotic and abiotic factors, such as the climate specific to the regions of origin of the samples, geographical factors such as altitude and the nature of the soil that direct the biosynthesis towards the preferential formation of specific products.

2.2. Variation of the majority compounds of *Eucalyptus salubris* according to the two harvest regions and the different extraction methods

The analysis of figure 15 shows that the compounds P-cymene, sabinene, and cryptone, are specific to the harvest of El-Hamma arboretum while the harvest of Zrig arboretum is characterized by 1,8-cineole. Concerning the El-Hamma harvest, the same majority compounds are found for the different extraction processes. However, for the harvest of Zrig, 8-Terpinene was present with simple hydrodistillation and microwave assisted extraction.

In this context, Ben Marzoug et *al.* (2010) reported that the analysis of the chemical composition of the essential oil of *Eucalyptussalubris* from South Tunisia obtained by HS identified 27 compounds representing 99.2% of the total. The major compounds were 1,8-cineole (71.3%) followed by trans-pino-carveol (6.0%), p-cymene (5.4%) and a-pinene (4.6%). Furthermore, Elaissi et *al.* (2006) reported that the main compounds of *E. salubris* collected from central Tunisia were a-pinene (11.1%), limonene (19.5%), 1,8-cineole (26.5%) and cryptone (11%).

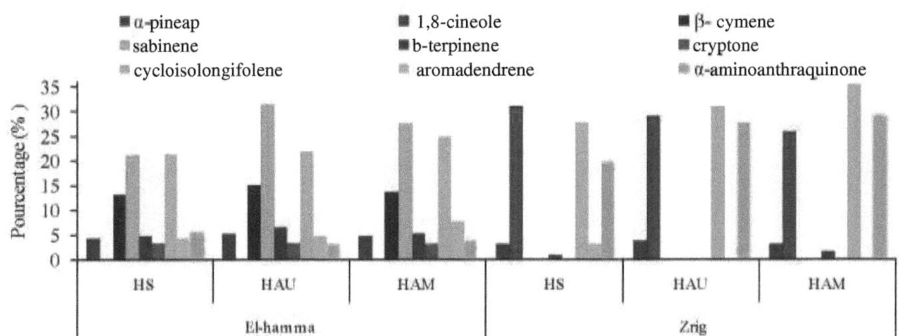

Figure 15: Variations in the percentages of major compounds in the essential oils of E. salubris from the two regions (El-hamma and Zrig) and extracted by three extraction methods

2. 3. Classification of essential oils by Principal Component Analysis (PCA)

In order to obtain a statistical description of the different extraction methods, we submitted the results to a Principal Component Analysis (PCA). Thus, the plane defined by the first two axes represents 76.79% of the total variance of the population (Axis 1: 55.97% and Axis 2: 20.82), indicating the presence of three distinct groups (Figure 16). Group I gathers the essential oils of *E. salubris* from El-hamma extracted by three extraction methods (HS, EAU and EAM). Their compositions are characterized by the presence of a-pinene (4.08-5.1-4.78%), p-cymene (13.05-14.95%), sabinene (21.15-31.35%), b-Terpinene (4.77-6.34%), Aromadendrene (4.2-7.56%) and Cryptone (2.85-3.3%) as the majority compounds.

Group II gathers the essential oils of *E. salubris* de Zrig extracted by the two modern methods (EAU and EAM); their compositions are characterized by the presence of a-Aminoanthraquinone (27.46-28.93%), 1,8-Cineole (28.99-25.69%), Cycloisolongifolene (30.85-35.27%), Borneole (2.26-2.29%) and p-Mentha-1,3,8-triene (4.83- 3.57%) as the majority compounds.

Group III gathers the essential oil of *E. salubris* de Zrig extracted by the classical extraction method (hydrodistillation), its composition is characterized by the presence of 2-Carene (5,01%), a-Phellandreneepoxide (5,35%) and Pinocarvone (2,26%), as major compounds.

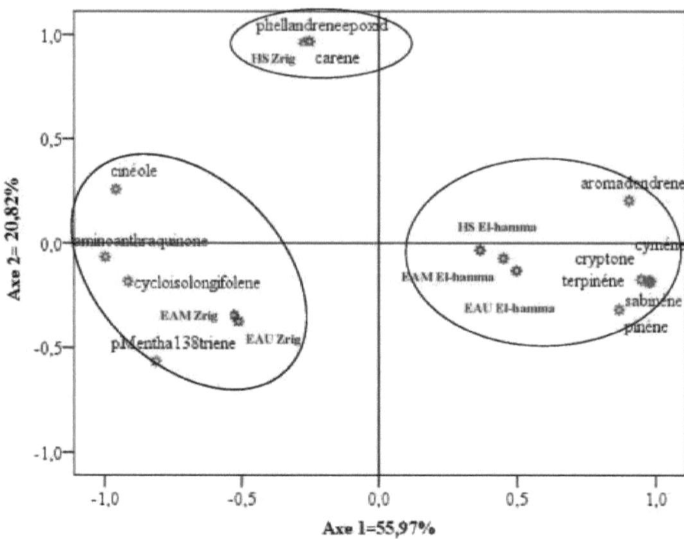

Figure 16: Distribution of essential oils of E. salubris from the two regions and extracted by three methods and their majority compounds according to the planes defined by axis 1 and axis 2 of the PCA

Our results showed the presence of quantitative and qualitative differences in the essential oils of *E. salubris* from both regions. Previous work has shown that oil biosynthesis is influenced by various environmental factors namely light intensity and climatic conditions (Li et *al.* , 1996). Similarly, Boland etal. (1991) indicated that within the genus *Eucalyptus*, the variation in the chemical composition of essential oils can be attributed to three major factors: leaf age, genetic factors and climatic factors. Furthermore, Batish et *al.* (2008) demonstrated that 1,8-cineole is the majority compound of the species of the genus *Eucalyptus*. According to Ben Marzoug et *al.* (2010), autumn harvests of *E. salubris* (El-Hamma, Gabes) revealed the richness of this oil in monoterpenes such as 1,8-cineole (71.3%), trans-pinocarveol (6%), p- Cymene (5.4%) and a-pinene (4.6%) On the other hand, Silou et *al.* (2009) showed that using steam stripping techniques and hydrodistillation, the intensity of heating and extraction time influence the yield and composition of essential oil of *E. cinerea*. Indeed, the best essential oil content is obtained by steam distillation after an extraction time of 90 min and a heating of 450 W giving a percentage of citronellal of 84.6%. A shorter extraction time (30 min) and a lower heating intensity (360 W) decrease the essential oil yield but slightly increase their citronellal content (85.4%). In addition, the work of Haouel (2017) showed that the composition of *E. salubris* is characterized byhigh contents of hydrocarbon monoterpenes varying from 51.5% for Simple Hydrodistillation (HS); 53.93% for Microwave Assisted Extraction (MAE) and 60.81% for Ultra Sound Assisted Extraction (USE). The hydrocarbon sesquiterpenes occupy the second place with contents varying from 30.61% (EAU) to 35.87% (EAM).

3. Evaluation of the insecticidal activity of essential oils

3.1. Regional variability in insecticidal activity

The results related to the evaluation of the insecticidal fumigation activity of *E. salubris* essential oils from the two regions (El-hamma and Zrig) extracted by three extraction methods against *S. oryzae* adults are shown in Figure 17. Examination of the figure showed that the insecticidal activity by fumigation varied with concentration, exposure time, harvest region and extraction method. Indeed, as the exposure time increases, the percentage of adult mortality increases. The examination of the results indicates that the essential oil of *E. salubris* obtained by simple hydrodistillation of El-Hamma presents a great effectiveness against the adults of *S. oryzae*. Indeed, after 12 h of exposure, the highest dose of 100 |^l/l air caused 100% mortality. In contrast, the essential oil of Zrig requires 66 h of exposure to achieve 100% mortality. Regarding the lowest concentration 16.66 |il/l air, the percentage of mortality reached 100% after 180 h of exposure with the essential oil of El-hamma while those of Zrig after 132 h of exposure reached 100% mortality.

For *E. salubris* essential oils obtained by ultrasound assisted extraction, comparison of the two harvests of El-hamma and Zrig showed that for all concentrations tested, the percentage of mortality of *S. oryzae* was always higher under the effect of El-hamma essential oil.
Concerning, the microwave assisted extraction, the comparison of El-hamma and Zrig oils showed that for the two low concentrations 16.67 and 33.33 |^l/l air, the percentage of mortality of Zrig essential oil is more effective than that of El-hamma contrary to the two high concentrations 50 and 100 |il/l air where El-hamma oil has a more important insecticidal effect than that of Zrig.

In sum, it can be stated that taking into account all the extraction processes, the oil from El-hamma arboretum is more effective than that from Zrig. This could be due to the higher contents of hydrocarbon monoterpenes (which vary from 49.24-62.29) and which are reported to be involved in insecticidal activities of essential oils.

Simple

Simple

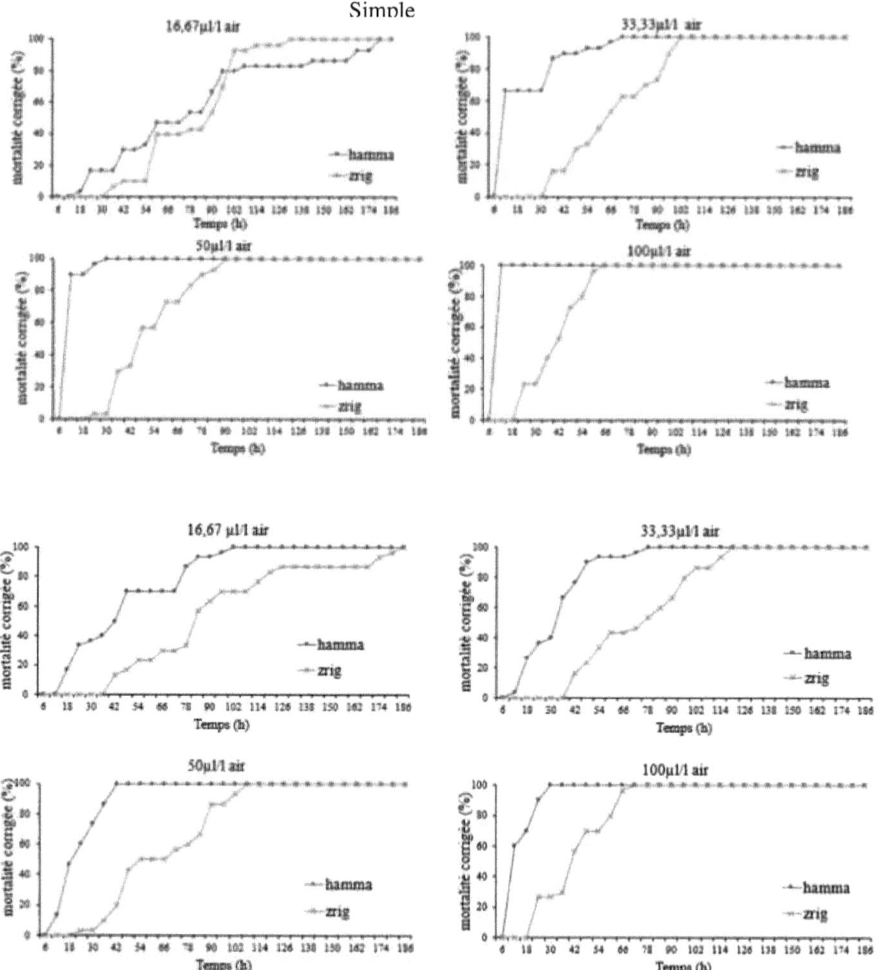

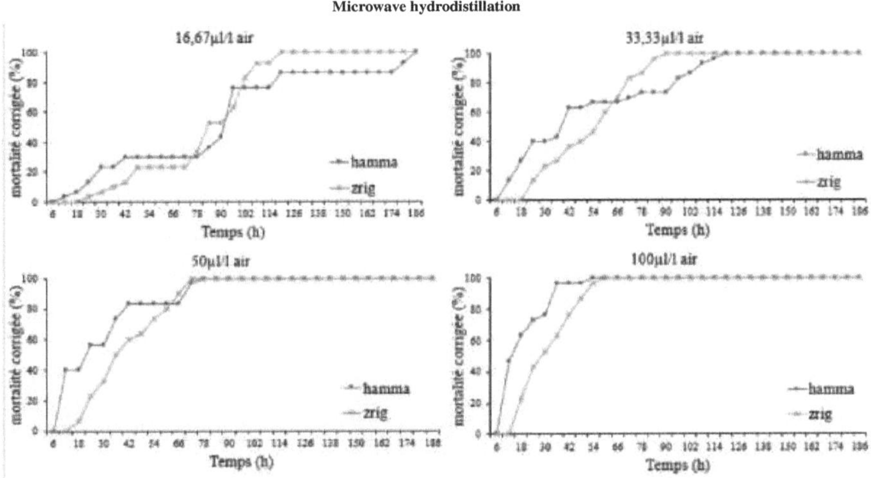

Figure 17: Evolution of corrected mortality (%) of Sitophilus oryzae adults according to different concentrations of Eucalyptus salubris essential oils from the two regions and obtained by three extraction methods.

3.2. Variability according to extraction methods

Figure 18 shows the corrected *S. oryzae* adult mortality for the four concentrations of *E. salubris* essential oils from the two regions and extracted using the different extraction processes. Examination of this figure shows that the insecticidal activity by fumigation varies according to the region of harvest, the extraction processes and the concentrations. First, it can be seen that regardless of the collection region and extraction procedures, all four concentrations tested caused total mortality of *S. oryzae* adults with varying exposure times.

Concerning the *E. salubris* oil of El-hamma, it is noticed that for the low concentration 16,67^l/l air the total mortality of 100% was reached after 102h for HS, against 180h and 186h for the modern extraction processes EAU and EAM respectively. And for the high concentration 100 |il/l air, 100% total mortality was reached after 12 h for HS, against 30 h and 54 h for WATER and EAM extraction processes respectively. Thus, the best process was the one that gave complete mortality in a shorter time. The processes are then ranked: HS > WATER > MAE.

Concerning the *E. salubris* oil from Zrig, it is noted that for the low concentration 16.67^l/l air, total mortality was obtained after 132h for HS, 186h for EAU and 120h for EAM and for the high concentration 100 |il/l air, total mortality was obtained after 66 44 h for HS, 70 h for EAU and 60 h for EAM. So, the best extraction process is EAM followed by HS and then by EAU.

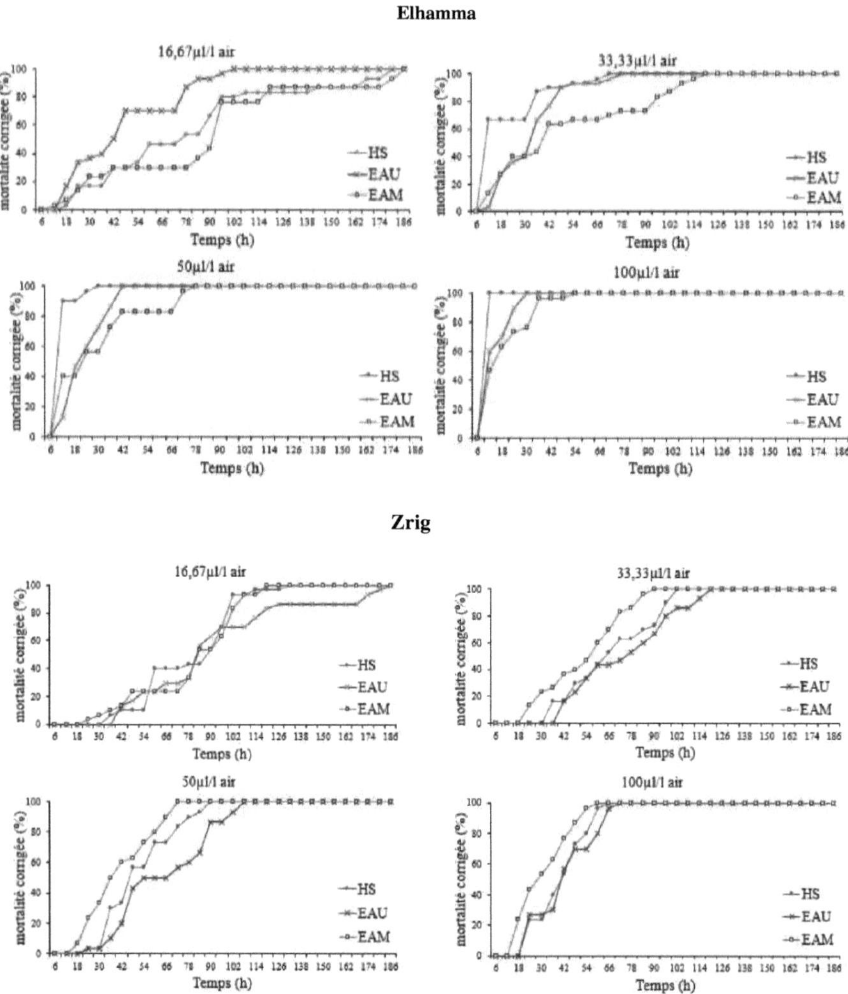

Figure 18: Evolution of corrected mortality (%) as a function of time of Sitophilus oryzae adults according to different concentrations of Eucalyptus salubris essential oils from the two regions and obtained by three extraction methods.

3.3. Variability by concentration

The evaluation of the insecticidal activity by fumigation of *E. salubris* essential oil from two regions (Zrig and El-hamma) and extracted by three processes simple hydrodistillation, microwave assisted extraction and ultrasound assisted extraction is expressed by the curves of corrected mortalities of *Sitophilus oryzae* adults according to the different durations of exposures to increasing concentrations of essential oil (Figure 19).

The results of the bioassays showed that fumigation toxicity varies with harvest region, extraction process, oil concentration, and duration or time of exposure. In addition, examination of Figure 19 indicates that for all three extraction processes, and for both harvest regions, as the concentration increases so does the percentage of mortalities is increasing.

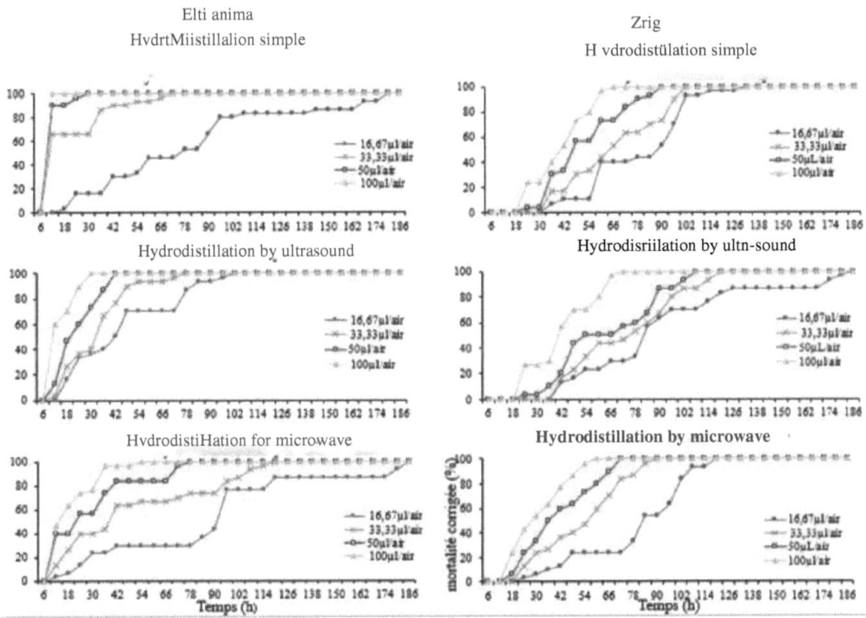

Figure 19: Evolution of corrected mortality (%) of Sitophilus oryzae adults according to the concentrations of Eucalyptus salubris essential oil obtained by three extraction processes.

These insecticidal activities probably have a close relationship with the chemical composition of essential oils as indicated by the results of several previous scientific works. The insecticidal potential of essential oils is attributed to their richness in bioactive compounds mainly monoterpenes (Batish et *al.*, 2008).

According to Woong et *al,* (2013), the insecticidal activity of a-pinene is more important than that of 1,8-cineole. a-pinene has a very high insecticidal activity compared with other constituents (Yeom et *al*, 2012). However, it would be difficult to think that the insecticidal activity of the essential oil is limited only to some of these majority constituents or to a specific compound, it could also be due to some minority constituents or to a synergistic effect of several constituents (Batish et *al.*, 2008). All this information has contributed to the understanding that the insecticidal effects of plant extracts depend on their chemical composition and the level of susceptibility of the insects tested (Isman, 2000).

These activities probably have a close relationship with the chemical composition of essential oils as indicated by the results of several scientific works. Obeng- Ofori et *al*,(1998) reported that the insecticidal activity of an essential oil is related to its chemical composition. Similarly Regnaualt-Roger and Hamraoui (1995) also indicated that oxygenated monoterpenes are more toxic against the adults of *Acanthoscelides obtectus* (Say). In addition, Batish et *al* (2008) indicated that the insecticidal activity of *Eucalyptus* essential oils *is* due to the presence and interaction of various compounds including 1,8-cineole, a-pinene, a-terpineol, aromadendrene and several other compounds.

The results of Haouel (2017) showed that the insecticidal activity by fumigation varies according to the extraction process, the harvest date, the concentration and the exposure time. Indeed, as the concentration increases, the percentage of mortality increases. For *Callosobruchus maculatus* adults, *E. salubris* essential oils extracted by ultrasound assisted hydrodistillation (UAS) and microwave assisted hydrodistillation (MAH) were remarkably distinguished by their faster fumigant effects compared to simple hydrodistillation (HS) for all four concentrations tested.

3.4. Hierarchical classification

The hierarchical classification performed from the data set based on the Euclidean distance between the different essential oils of the two arboreta and extracted by three methods allowed the identification of 2 groups of species (A and B) with a dissimilarity d>10 (Figure 20).

The group analysis, carried out on the basis of the average mortality following the action of the essential oils of *E. salubris* from the two regions and extracted by three extraction processes, revealed little variability and allowed the differentiation of two groups A and B (Figure 20).

- **Group A** is represented by the oils from the El-hamma region which are characterized by a very remarkable efficiency against the adults of *S. oryzae*.
- **Group B** is represented by the oils from the Zrig region which are characterized by a lower efficacy than the oils harvested from El-hamma against *S. oryzae* adults.

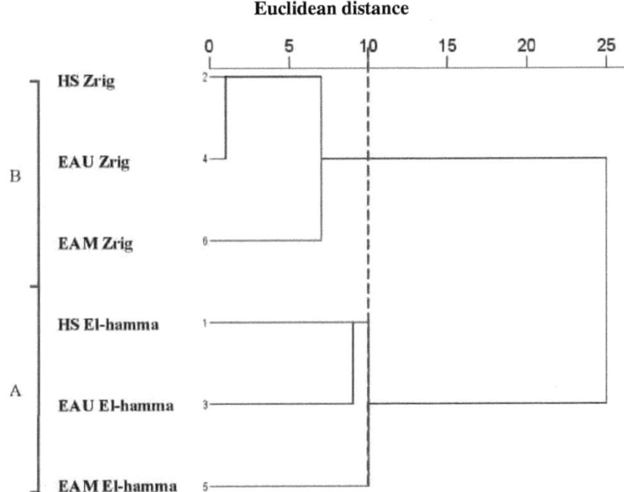

Figure 20: Similarity dendrogram from the cluster analysis based on the average mortality caused by essential oils

The correlation matrix (Table 4), presents the correlation coefficients between *S. oryzae* adult mortality, harvest regions, extraction methods, concentration and exposure time, taken 2 to 2. Observation of Table 4, shows that the correlation coefficients between mortality and harvest regions are highly significant negative (-0.40). While the correlation coefficients between mortality and concentration and exposure time are highly significant positive (0.57 and 0.52).

Table 5: Matrix of correlations between Sitophilus oryzae adult mortality, harvest regions, extraction methods, concentration and exposure time.

	Harvesting regions	Method extraction	Concentration	Exposure time
Mortality	** -0,40**	-0,11	** 0,57**	** 0,52**

**. The correlation is significant at the 0.01 level (two-tailed).

The work of Bousbia, (2011) showed that the plant material subjected to hydrodistillation appears to be very similar to that of the untreated plant material. This author also reported that the changes observed for EAM extraction are markedly different from those observed by HS, clearly showing that the cells are broken or damaged during microwave treatment. This indicates that the mechanical stresses induced by the rapid decompression and violent vaporization of water have two main effects the dehydration effect and a subsequent change in the surface pressure of the glandular wall, causing the cell to deteriorate or even burst. Similar effects were reported by Paré and Bélanger (1997) who indicated that during microwave-assisted extraction of rosemary leaves, the glands were subjected to more severe heat stress and high localized pressures, as in microwave heating, a pressure build-up in the glands may have exceeded their expansion capacity and caused a more rapid rupture compared to conventional extraction. Thus, microwaves allow a faster release of the essential oil contained in the plant matrix due to the almost instantaneous opening of the secretory glands. During a MAE or HS, the quantity of aromatic molecules, oxygenated or not, is clearly lower than the quantity of water present in the plant matrix. The boiling temperature of the mixture (water plus aromatic compounds) is therefore imposed by the boiling temperature of the water, i.e. 100°C, and does not depend in any way on the boiling temperature of the essential oil compounds. The modern processes are distinguished by their faster fumigant effects compared to simple hydrodistillation, they are environmentally friendly extraction processes and are suitable for sample preparation for the determination of essential oils in aromatic matrices or for their production. The reduction of extraction cost by these processes is highlighted in terms of energy, solvent consumption and time required.

3.5. Determination of lethal concentrations LC50 and LC95

The results of the calculation of the LC50 and LC95 of the essential oils of *E. salubris* extracted by the three extraction processes and tested against *S. oryzae* adults are shown in Table 6.

Table 6: Values of lethal LC50 and LC95 (gl/l air) concentrations of adult Sitophilus oryzae subjected to Eucalyptus salubris essential oils.

Region	Extraction process	LC50	CL95
ArboretumEl-hamma	Hydrodistillation simple	25,96 (22,142-29,730)	51,92 (43,104-71,281)
	Assisted extraction by ultrasound	34,433 (25,002-44,853)	202,197 (118,578-701,244)
	Assisted extraction by microwave	46,737 (36,045-63,143)	260,927 (147,226-973,244)
Zrig Arboretum	Hydrodistillation simple	149.378 (109.549-599.212)	376,175 (196,923-11248,770)
	Assisted extraction by ultrasound	137,754 (105,390-374,828)	325,891 (185,341-4158,711)
	Assisted extraction by microwave	117,090 (80,422-304,999)	725,197 (286,844-12126,153)

Examination of the lethal concentration results shows that the LC50 and LC95 values of the oils from El-hamma arboretum from simple hydrodistillation is lower than those from ultrasound assisted extraction and microwave assisted extraction. The order of insecticidal toxicity is as follows HS > WATER > MAE.

For the Zrig region, examination of the table showed that the LC50 values of microwave assisted extraction is lower than ultrasound assisted extraction and simple hydrodistillation. The order of insecticidal toxicity is as follows MAE > WATER > HS.

Examination of the results relating to lethal concentrations shows that for all extraction processes (HS, EAU, EAM), the LC50 and LC95 values for El-Hamma oils are significantly lower than those obtained for the oils from the Zrig arboretum; this demonstrates that the El-Hamma oils are more effective and more toxic against *S. oryzae* adults.

The work of Haouel (2017) showed that fumigation with *Eucalyptus salubris* essential oil against *C. maculatus* adults gave variable LC50 values depending on the extraction process (HS 21.39 pl/l air, EAU 13.47 pl/l air; EAM 16.58 pl/l air).

Our present results show that there is a great interest in developing *Eucalyptus salubris* essential oil fumigation as an alternative to industrial fumigant gas for the control of stored food pests.

4. Evaluation of the antioxidant activity of essential oils

The results of the measurement of antioxidant activity of *E. salubris* essential oils extracted by the three extraction processes and from the two arboretums El-hamma and Zrig are shown in Figure 21.

Figure 21: Percentage of inhibition of Eucalyptus salubris oil from two collection regions and obtained by three extraction processes.

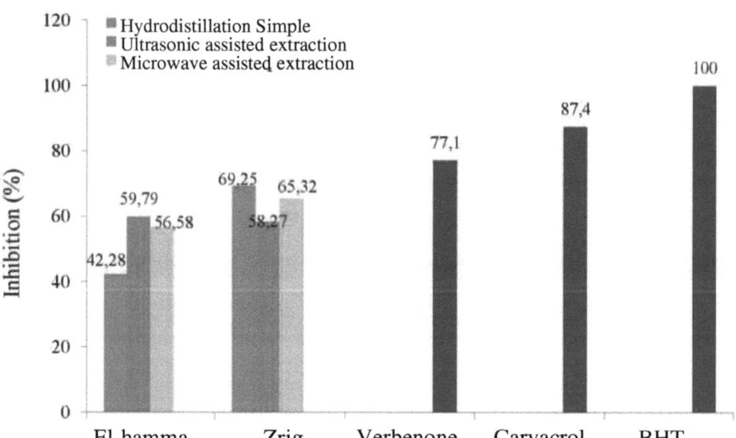

Examination of Figure 21 shows that the percentage of inhibition (PI) of the free radical (DPPH) varies according to the extraction process of the essential oil and the region of harvest. Indeed, for the region of El-hamma, the percentages of inhibition for the essential oil obtained by Extraction assisted by ultrasound were higher than those of the percentages of extraction assisted by microwave and those of the percentages of simple hydrodistillation. Indeed, the IP of El-Hamma essential oil obtained by the new extraction techniques have almost the same IP (WATER:59.79%) and (MAE:56.58%). This shows that the new processes have a higher antioxidant power than the old process (HS).

For the Zrig region, unlike the essential oil of El-Hamma, extraction by HS has a higher antioxidant activity than that obtained by EAU and EAM which have almost the same activity.

The essential oil of *Eucalyptus salubris* of Zrig obtained by simple hydrodistillation has a similar activity to that obtained by the synthetic antioxidant Verbenone and far from carvacrol and BHT.

The work of Naceur Ben Marzoug et al (2010), showed that the antioxidant activity obtained by the DDPH test of the essential oil of *Eucalyptus salubris* is weaker than that of *E. gracilis* and the essential oil of *E. salmonophloia* but more important than the essential oil of *E. oleosa*

These activities probably have a close relationship with the chemical composition of EOs as indicated by the results of several scientific works. In this context, Achour (2012) already proved that the antioxidant power of *E. cinerea* essences is mainly due to its chemical profile rich in phenolic compounds such as carvacrol as well as other oxygenated and hydrocarbon components like 1,8-cineole, a- terpinolen, 8-terpinene. On the other hand, Laib (2012) proved that the antioxidant activity is not only due to the majority compounds, but there may be other compounds that can interact in a synergistic or antagonistic way to create an effective system towards free radicals. Previous work has shown that the incorporation of essential oils directly into food (ground meat, fruit puree, yoghurt, etc. ...) or their application by spraying the surface of the food helps to preserve the food from oxidation phenomena (Caillet and Lacroix, 2007).

In the light of all these data illustrated and which approve the antioxidant capacity that the essential oils of *Eucalyptus* possess, they can be studied for their possible use as an alternative for the protection of food against oxidation.

General conclusion

The present work is part of the research program on alternatives to fumigation by synthetic chemicals through products of natural origin. In this work, we undertook the study of the insecticidal and antioxidant properties of essential oils of *Eucalyptus salubris* of arboreta of El-hamma (Gabes) and Zrig (Gabes) against one of the main pests of stored goods the rice weevil *Sitophilus oryzae*.

The first part of the work focused on the comparison of the effect of modern extraction techniques (Extraction assisted by ultrasound EAU and Extraction assisted by microwave EAM) and conventional (Hydrodistillation simple HS) on the yield and chemical composition of essential oils of arboreta of El-hamma (Gabes) and Zrig (Gabes). The results obtained showed that these oils presented interesting yields. The highest yield was observed for essential oils extracted by modern methods (EAU, EAM). In addition, the results obtained showed that the order of ranking of extraction processes is as follows: microwave assisted hydrodistillation>ultra-sound assisted hydrodistillation>simple hydrodistillation.

In addition, the chemical analyses indicated that the essential oils of El-hamma are characterized by the highest percentages of monoterpenes, regardless of the extraction process. Indeed, these proportions were 50.97% for HS, 56.95% for EAM and 63.13% for EAU. However, the essential oils of Zrig arboretum are marked by high contents of a-aminoanthraquinone (19.71% for HS, 27.46% for EAU and 28.93% for EAM) and cycloisolongifolene (27.87 for HS, 30.85 for EAU and 35.27% for EAM) obtained by the different extraction processes (HS, EAU, EAM). In addition, these oils were characterized by high percentages of oxygenated monoterpenes (27.78% for EAM, 31.25% for EAU and 39.59% for HS) and represented mainly by 1,8-cineole.

The second part of the work focused on determining the effect of extraction methods on the insecticidal activity by fumigation against *S. oryzae* adults. The results obtained proved the insecticidal potential of the different essential oils against *S. oryzae*. The essential oils of El-hamma gave the best insecticidal activities compared to those of Zrig. The conventional HS method gave the best performance for both arboreta in terms of mortality and lethal concentrations. For modern methods, ultrasound assisted extraction (UEE) was better for El-hamma oils (34.433^1/1 air) than Zrig (137.754^1/1 air). In contrast, microwave-assisted extraction (MAE) was better for Zrig oils (117.090^1/1 air) than El-hamma (46.737^1/1 air).

The last part of this study was devoted to the study of the antioxidant activity of essential oils of El-hamma and Zrig. The results showed that the six *Eucalyptus salubris* oils tested have a significant antioxidant activity that varies with the region of collection and the extraction process. The highest inhibitions were obtained for the Zrig oil with the following ranking: HS (69.25%) >

EAM (65.32%)> WATER (58.27%). In contrast, for the oils harvested from El-hamma, the modern processes WATER and MAE had the highest inhibition percentages (59.79% and 56.58% respectively) compared to the conventional HS method (42.28%).

In the light of these results, the activities of these oils deserve to be studied in greater depth under real conditions of storage warehouses in order to develop new alternatives for the control of stored food pests through the investigation of their toxic effect, which could improve the profitability of this control approach. Indeed, the development of bio-insecticides from extracts and essential oils of plants is part of the framework of agriculture and sustainable development. In addition, we propose that its insecticidal activities be studied on the first stages of the crop, which are very harmful to stored foodstuffs. We also suggest that the insecticidal potential of these days be tested on other pests.

Bibliographic References

Achour N. K., 2012. Study of the chemical composition of essences of four species of Eucalyptus growing in the region of Tizi Ouzou. Thesis magister. Faculty of Science. Scientific university Mouloud Mameri-Tizi Ouzou.

Anastas P.T., Warner J., John C., 1998. Green chemistry: theory and practice. Oxford: Oxford University. Press ,2000.

Anonymous 4, 2008. International Atomic Energy Agency. Potential markets for sterile date.moth for use in SIT in Tunisia. *In*: Model business plan for a sterile insect production facility, Annex 5, 313-320 pp.

Aragon S., 2014. Weevil in Encyclopaedia Universalis.

Armitage D.M.,1987. Controlling insects by cooling grain. BCPC Mono.No. 37, Stored product.

Asbahani A, Miladi K, Badri W, Sala M, Aït Addi EH, Casabianca H, El Mousadik A, Hartmann D, Jilale A, Renaud FNR, Elaissari A., 2015. Essential oils: from extraction to encapsulation. International Journal of Pharmaceutics 483, 220-243.

Azmir J., Zaidul I.S.M., Rahman M.M., Sharif K.M., Mohamed A., Sahena F., Jahurul M.H.A., Ghafoor K., Norulaini N.A.N., Omar A.K.M., 2013. Techniques for Extraction of Bioactive Compounds from Plant Materials: A Review, Journal of Food Engineering, 117, p426-436.

Bachta M.S., 2011. Food subsidies amounted in 2011 to 1150 million dinars or more than 6% of total public expenditure.

Bajji M., 1999. Study of resistance mechanisms to water stress in durum wheat: characterization of cultivars differing in their levels of drought resistance and somaclonal variants selected in vitro. D. thesis, Faculty of Science, Catholic University of Leuven, 1999.

Bakkali F., Averbeck S., Averbeck D., Idaomar M., 2008. Biological effects of essential oils-a review. Food Chem. Toxicol. 46, p446-475.

Balachowsky A.S., 1963. Entomology applied to agriculture. Traité, Tome 1 coléoptères. Ed. Masson et Cie. Paris. Pp: 874-1263.

Bandoniene D., Pukalskas A., Venskutonis P. R., and Gruzdiene D., (2000). Preliminary screening of antioxidant activity of some plant extracts in rapeseed oil. Food Research International, 33, p785-791.

Batish D.R., Singh P.H., Kohli K.R., Kaur S., 2008. *Eucalyptus* essential oil as a natural pesticide. Forest Ecol. Manag. 256: p2166-2174.

Belyagoubi L., 2005. Effect of some plant species on the growth of cereal spoilage fungi. Magistère. Université Abou Bakr Belkaid. P: 61.

Ben Hassine D., 2008. Study of the chemical composition and antimicrobial activity of four essential oils of *Eucalyptus*. Master thesis, National Agronomic Institute, Tunisia, p83.

Benamor B., 2008 Maitrise de l'Aptitude Technologique de la Matière Végétale dans les Opérations d'Extraction de Principes Actifs ; Texturation par Détente Instantanée Contrôlée DIC. PhD thesis, University of La Rochelle, France.

Benayad N., 2008. Laboratory of natural substances and flash thermolysis. Departempent de chimie.

Faculté des sciences de Rabat. Essential oils extracted from Moroccan medicinal plants: an effective means of pest control in stored foodstuffs 2008.

Benazzeddine S., 2010. Insecticidal activity of five essential oils against *Sitophilus oryzae* (Coleoptera; Curculionidae) and *Tribolium confusum* (Coleoptera; Tenebrionidae) Ecole nationale supérieure agronomique El Harrach Algeria - State Engineer in Agronomic Sciences 2010.

Ben Marzoug N.H., Bouajila J., Ennajar M., Lebrihi A., Florence M., Couderc F., Abderraba M., and Romdhane M., **2010**.*Eucalyptus* (*gracilis,oleosa, salubris, and salmonophloia*) Essential Oils:Their Chemical Composition and Antioxidant and Antimicrobial Activities. *J Med Food* 13 (4), 1-8.

Boland DJ, Brophy JJ, House APN (1991). Eucalyptus Leaf Oils: Use, Chemistry, Distillation and Marketing. Inkata Press. Melbourne.

Bousbia N., 2011. Extraction of antioxidant-rich essential oils from natural products and agri-food co-products. Other. Universite d'Avignon; Institut national agronomique (El Harrach, Algerie), 2011. French.<NNT: 2011AVIG0243>.

Boutamani M., 2013.Study of the variation of yield and chemical composition of *Curcuma longa* and *Myristica fragrans as a* function of time and technique used. University of science and technology Houari Boumediene Algiers - Master domain chemistry of medicine 2013.

Brooker M.I.H., Kleinig D.A., 2006. Field guide to Eucalyptus ([2nd]). Northem Australia (Vol.3) Melbourne. Bloomings books.

Bruneton J., 1993. Pharmacognosy, Phytochemistry, Medicinal Plants. Paris: éditions médicales internationales. Editions Tec et Doc Lavoisier.409-417.

Caillet S. & Lacroix M., 2007. "Essential oils: their antimicrobial properties and potential applications in food", *INRS -Institut Armand-Frappier, (RESALA), 2007*, P: 1 - 8.

Cimanga K., Kambu K., Tona L., Apers S., De Bruyne T., Hermans N., Totté J., Pieters L., Vlietinck A. J., 2002. Correlation between chemical composition and antibacterialactivity of essential oils of some aromatic medicinal plants growing in the Democratic Republic of Congo. J. Ethnopharmacol. 79, p213-220.

Cheikh M'hamed H. Evaluation of the tolerance of some barley accessions to salt stress. DEA, Institut national agronomique deTunisie (Inat), 2004.

Clifford A.A., 2002. Extraction of Natural Products with Superheated Water. In: Clark J., Macquarrie D. (Eds) Handbook of Green Chemistry and Technology. Blackwell Science Ltd. p524-531.

Codon and William, 1991. Insecticidal activity of five essential oils against *Sitophilus oryzae* (Coleoptera; Curculionidae) and *Tribolium confusum* (Coleoptera; Tenebrionidae).

Cruz J F and Troude F., 1988. Conservation of grains in hot regions. Collection of the Ministry of Cooperation and Development. Techniques rurales en Afriques. CEEMAT/CIRAD, Montpellier, 548p.

De Castro L., Capote F. 2007 Ultrasound-assisted preparation of liquid samples. TalantaVolume 72, Issue 2: 321-334.

De Groot I (2004). Protection of stored cereals and legumes. Agrodok 18, Agromisa Foundation. Wageningen, p74.

Duke J.A.,2004. Dr. Duke's phytochemical and ethnobotanical databases. Available online at http://www.ars-grin.gov/duke/(accessed on 9 june,2008).

Egbon I.N and Ayertey J.N., 2013. Incidence of *Sitophilus oryzae* and Other Stored-product Pests on Cowpea in Local Markets in Accra: Management Strategies Employed by Retailers. Pakistan Journal of Biological Sciences, 16: p435-438.

Elaissi A, Medini H, Chraief I, Marzouk H, Bannour F, Farhat F,Ben Salah M, Chemli R, Khouja ML, 2006. Contribution to the qualitative and quantitative study of eleven Eucalyptus species essential oil harvested of Hajeb's Layoun arboreta (Tunisia). In: Revue des Régions Arides, Special Issue, Proceedings of the International Seminar on Perfume, Aromatic and Medicinal Plants. International Center for Agricultural Research in Dry Areas, bAU1 Medenine, Tunisia, 2006, pp. 173-177.

El haib A., 2011. Valorization of natural terpenes from Moroccan plants by catalytic transformations. PhD thesis, Toulouse III - Paul Sabatier University, 158.

El Kalamouni C., 2010. Chemical and biological characterization of extracts of forgotten aromatic plants from Midi-Pyrénées. PhD thesis in Agroresources Sciences University of Toulouse: 61-62.

Evans W.C., 1998 - Trease and Evan's Pharmacognosy, [14th] edition SANDERS, pp. 48- 65, 612.

Fiori A.C.G., Schwan-Estrada K.R.F., Stangarlin J.R., Vida J.B., Scapim CA, Cruz M.E.S., Pascholati S.F., 2000. Antifungal activity of leaf extracts and essential oils of some medicinal plants against *Didymella bryoniae*. J. phytopathol. 148, p483-487.

Flemming, S., 1997. The herb book. How to grow them, identify them and use them in cooking. Ed Chantecler, Belgium, 116.

Frankel E. N., Meyer A S., 2000. The problems of using one-dimensional methods to evaluate multidimensional food and biological antioxidants, *Journal of Science and Food Agriculture*, 80: p1925-1941.

Gawde A., Cantrell C.L., Zheljazkov V.D, Astatkie T and Schlegel V. 2014 Steam distillation extraction kinetics regression models to predict essential oil yield, composition, and bioactivity of chamomile oil", Industrial Crops and Products (58): 61-67.

Garneau F.X., 2001. Notes from the Natural Products course. Department of Science UQAC, Chicoutimi, Quebec. P17.

Gérin M., 2002 Solvents and Prevention: New Perspectives. In: Gérin M. (Ed) SolvantsIndustriels: Santé, Sécurité, Substitution. Masson, Paris. P1-12.

Haouel S., 2017. Unpublished doctoral thesis, Faculty of Sciences of Tunis University of Tunis El Manar.

Hebert J. P., Troude F., Griffon D., and Cruz J.F., 1972. Ministry of Cooperation and Development. Centre d'études et d'expérimentation du machinisme agricole et tropical. Paris, France. p400-424.

Henry Y. S., Andrew B., Archana G., Charles L. C, Tess A., Augustine E. O., Valtcho D. Z., Vicki S., 2014. Hydrodistillation time affects dill seed essential oil yield, composition, and bioactivity in Industrial Crops and Products.

Hernandez Ochoa L. R., 2005. Substitution of synthetic solvents and active materials by a "solvent/active" combination of plant origin. PhD thesis, Ecole Nationale Supérieure des Ingénieurs en Arts Chimiques et Technologiques, Toulouse France.

Isman M.B., 2000. Plant essential oils for pest and disease management. Crop protection. 19: pp. 603-608.

Jacobs M.R. 1982. *Eucalyptus* in reforestation pp428-441 and p658-661.

Jae-Seoun Hur JS, Ahn SY, Young Jin Koh YJ, Lee C.,2000. Antimicrobial properties of cold-tolerant Eucalyptus species against phytopathogenic fungi and food-borne bacterial pathogens. Plant Pathol J 2000; 16:286-289.

Jarraya A., 2003. Main pests of cultivated plants and stored goods in North Africa: their biology, natural enemies, damage and control. Climat Pub, Tunis. P415.

Khouja M.L., Khaldi A., Rejeb M.N., 2001. Results of the *Eucalyptus* introduction trials in Tunisia. Proceedings of the International Conference *Eucalyptus* in the Mediterranean basin: Perspectives and new utilization. Centro Promozione Pubblicita, Florence Italy, 163.

Khouja M.L., Souyah N., Ghrabi Z., Ben Haj Jilani I., Khmiri A., Khaldi A., 2006. Flowering calendar of *Eucalyptus* species in the melliferous arboretum of Sidi Ismaîl. Annales de l'INRGREF, 9, Special issue, p241-254.

Kranz J., Schmutterer H., Koch W., 1977. Diseases Pest and Weeds in tropical crops. Verlag Paul Parey. Berlin. p666.

Labuza T.P., 1971. Kinetics of lipid oxidation in foods; Critical Reviews in Food and Technology; P: 355-405.

Laib I., 2012. Study of antioxidant and antifungal activities of essential oil from dry flowers of *Lavandula officinalis*: application to moulds of pulses. *Journal "Nature & Technology"*. n° 07. Pages 44 to 52.

Leonelli C., Veronesi P., Cravotto G. ,   **2013.** Microwave-Assisted Extraction: An Introduction to Dielectric Heating. In: Chemat F., Cravotto G. (Eds.), Microwave-assisted Extraction for Bioactive Compounds: Theory and Practice. Springer. p1-14.

Lepesme P., 1944. The beetles of foodstuffs and stored industrial products. Paul Lechevallier, Paris .129-130P

Li H., Madden, J.L., Potts, B.M., 1996. Variation in volatile leaf oils of the oils of the Tasmanian Eucalyptus species II. Subgenus Symphyomytrus. Biochem. Syst. Ecol. 24, p547- 569.

Linnee, 1763. phasma necydaloides (linné,1763) philippe lelong le ferradoun 3,31570 saine foy d'aigrefeuille, France.

Mahtab S., Zurina Z. A., Robiah Y., Dayang R. A. B.,Hiroyuki Y., Eng H.L., 2017. Assessing the kinetic model of hydro-distillation and chemical composition of *Aquilaria malaccensis* leaves essential oil. Chinese Journal of Chemical Engineering vol 25, p216-22.

Mansouri N., Satrani B., Ghanmi M., El Ghadraoui L., Aafi A., 2011. Chemical and biological study of essential oils of *Juniperus phoenicea* ssp. *lycia* and *Juniperus phoenicea* ssp. *turbinata* from Morocco. *Biotechnol. Agron. Soc. Environ.* 2011 15(3), 415-424.

Markham R.H., Bosque-Pérez N.A., Borgemeister C., Meikle W.G., 1994. Developing pest management strategies for *Sitophilus zeamais* and *Prostephanus truncatus*. Tropics FAO plan prot.

42: 97-116.

Mediouni Ben Jemâa J., Haouel S., Bouaziz M., Khouja M.L., 2012. Seasonal variations in chemical composition and fumigant activity of five *Eucalyptus* essential oils against three moth pests of stored dates in Tunisia. Journal of Stored Products Research 48, 61R67.

Metro A.,1970. Eucalyptus in the Mediterranean world. Ed. Masson et Cie, Paris. P 513.

Mohammedi Z., 2006. Study of the antimicrobial and antioxidant power of essential oils and flavonoids of some plants of the Tlemcen region. Master's thesis, Abou Baker Belgaid University Tlémcen Algeria. 155 p.

Momar T. G., Dogo S., Wathelet J., Lognay G., 2010. Pest management of cereal and legume stocks in Senegal and West Africa: a literature review. BASE p 183-194.

Muhlen CV, Zini CA, Caramao EB, Marriott PJ. 2008. Comparative study of *Eucalyptus* dunnii volatile oil composition using retention indices and comprehensive two-dimensional gas chromatography coupled to time-of-flight and quadrupole mass spectrometry. J Chromatogr A 2008; 1200: p34-42.

Obeng-Ofori D., Reichmuth C.H., Bekele J., Hassanali A., 1998. Toxicity and protectant potential of camphor, a major component of essential oil of Ocimum kilimandscharicum against four product beetles - International Journal of Pest Management, vol. 44, 203-209 pp.

Oldrich L., Klejdus B., Ladislav K., Michaela D., Khaled A., Vlastimil K., Richard H. 2005,Biochemical systematics and Ecology 33, pp. 983-992.

Pare'J. R. J., Bélanger J. M. R., (1997). Microwave-Assisted Process (MAPTM): Principles and Applications. In J. R. J. Pare' & J. M. R. Be'langer (Eds.), Instrumental methods in food analysis (pp. 395-420). Amsterdam: Elsevier Science.

Pavela R.,2009. Larvicidal property of essential oils Culex quinquefasciauts Say 5 (Diptera: Culicidae) Industrial Corps and Products, 30(2009), pp.311-315.

Peyron L., 1992: Current classical techniques for the manufacture of natural aromatic raw materials. Chapter 10, pp 217 - 238. Cited in: Les arômes alimentaires. Coordinators RICHARD H. and MULTON J.-L. Ed. Tec& Doc-Lavoisier and Apria. p438.

Regnault-Roger C., Hamraoui A., 1995. Fumigant toxic activity and reproductive inhibition induced by monoterpenes on *Acanthoscelides obtectus* (Say) (coleoptera), a bruchid of kidney bean (Phaseolus vulgaris L.) Journal of Stored Products Research, Volume 31: 291- 299.

Regnault-Roger C., and Philogène B.J.R., 2005. Evolution of synthetic organic insecticides. In: Phytosanitary issues for agriculture and the environment; (eds.Renault- Roger,C., Fabres, G.,Philogène, B.J.R.). Edition TEC et DOC. Paris. P 20-43.

Saad E.Z., Hussein R., Saher F., Ahmed Z., 2006. Acaricidal activities of some essential oils and their monoterpenoidal constituents against house dust mite *Dermatophagoides pteronyssinus* (Acari:v Pyroglyphidae). J. Zhejiang Univ. Sci. B. 7, 957-962.

Sabbour.M.M., 2012. Entomotoxicity assay of two nanoparticle materials against *Sitophilus oryzae* under laboratory and store conditions in Egypt. J. Nov.Appl. Sci.1, 103-108.

Sefidkon F., Assareh M.H, Abravesh Z., Barazandeh M.M: 2007. Chemical composition of the essential oils of four cultivated Eucalyptus species in Iran as medicinal plants (*E. microtheca, E.*

spathulata, E. largiflorens and E. torquata). Iran J Pharm Res 2007;6: p135- 140.

Silou T., Loumouamou AN., Loukakou E., Chalchat JC., Figuérédo G., 2009. Intra and interspecific variations of yield and chemical composition of essential oils from five eucalyptus species growing in the Congo-Brazzaville Corymbia subgenus. J. Essent. Oil Res. 21(3):203 210.

Silvy, 1992. Insecticidal activity of five essential oils against *Sitophilus oryzae* (Coleoptera; curculionidae) and *Tribolium confusum* (Coleoptera; Tenebrionidae).

Svoboda K.P., Greenaway R.I. 2003. Investigation of volatile oil glands of *Satureja Hortensis L.* (Summer savory) and phytochemical comparison of different varieties. The International Journal of Aromatherapy. 13(4):196-202.

Tapondjou L.A., Adler C., Bouda H. & Fontem D.A., 2002, Efficacy of powder and essential oif from *Chenopodium ambrosioides* leaves as post-harvest grain protectants against six-stored product beetles. Journal of Stored Products Research, 38, 395-402.

Taylor R. W. D., Harris A. H., 1994. Control of the larger grain borer, Prostephanus truncates (Horn) (Coleoptera: Bostrichidae), in bagged maize by fumigation under gas proof sheets. FAO Plant protection in press.

Tietze L. F., Modi A., 2000. Multicomponent domino reactions for the synthesis of biologically active natural products and drugs, *Medicinal Research Reviews, 20* (4): p304- 322.

Troude F., Griffon D., Hébert J.P., Cruz J.F., 1988. Ministére de la coopération et du développement. Centre d'études et d'expérimentation du machinisme agricole et tropical. Paris, France.443-454p.

Tsiri D., Kretsi O., Chinou I.B., Spyropoulos C.G., 2003. Composition of fruit volatiles and annual changes in the volatiles of leaves of *Eucalyptus camaldulensis* Dehn. growing in Greece. Flavour and Fragrance Journal, 18(3): 244R247.

Va'zquez G, Fontenla E, Santos J, Freire MS, GonzOalez-Alvarez J, Antorrena G., 2008. Antioxidant activity and phenolic content of chestnut (Castanea sativa) shell and Eucalyptus (Eucalyptus globulus) bark extracts. Ind Crop Prod 2008;28: p279-285.

Yeom H.J., kang J.S., kim G.H., park I.k., 2012. Insecticidal and acetylcholine esterase inhibition activity of Apiceae plant essential oils and their constituents against adults of german cockroach (Blatella germamca). J. agnc. Food Chem.60, 7194-7203.

Woong-Sung K., Kang J., Park K.I., 2013. Fumigant toxicity of Apiceae essential oils and their constituents against Sitophilus oryzae and their acethylcholinesterase inhibitory activity.Journal of Asia -Pacific Entomology 16, 443-448.

Webography

Anonymous 1: http://www.ozanimals.com/Insect/Granary-Weevil/Sitophilus/granarius.html.

Anonymous 2: http://www.ozanimals.com/Insect/Greater-Rice-Weevil/Sitophilus/zeamais.html.

Anonymous 3: https://www.anbg.gov.au/cpbr/cd-keys/Euclid/sample/html/SALUBR.htm.

I want morebooks!

Buy your books fast and straightforward online - at one of world's fastest growing online book stores! Environmentally sound due to Print-on-Demand technologies.

Buy your books online at
www.morebooks.shop

Kaufen Sie Ihre Bücher schnell und unkompliziert online – auf einer der am schnellsten wachsenden Buchhandelsplattformen weltweit! Dank Print-On-Demand umwelt- und ressourcenschonend produziert.

Bücher schneller online kaufen
www.morebooks.shop

Printed by Books on Demand GmbH, Norderstedt / Germany